VoLTE and ViLTE

VoLTE and ViLTE

Voice and Conversational Video Services over the 4G Mobile Network

André Perez

iSTE

WILEY

First published 2016 in Great Britain and the United States by ISTE Ltd and John Wiley & Sons, Inc.

ISTE Ltd
27-37 St George's Road
London SW19 4EU
UK

www.iste.co.uk

John Wiley & Sons, Inc.
111 River Street
Hoboken, NJ 07030
USA

www.wiley.com

Library of Congress Control Number: 2016938934

British Library Cataloguing-in-Publication Data
A CIP record for this book is available from the British Library
ISBN 978-1-84821-923-6

Contents

Preface

This book presents the mechanisms used in the 4G evolved packet system (EPS) mobile network and in the IP Multimedia sub-system (IMS) for the supply of voice over long term evolution (VoLTE) and video over long term evolution (ViLTE) service (Figure 1).

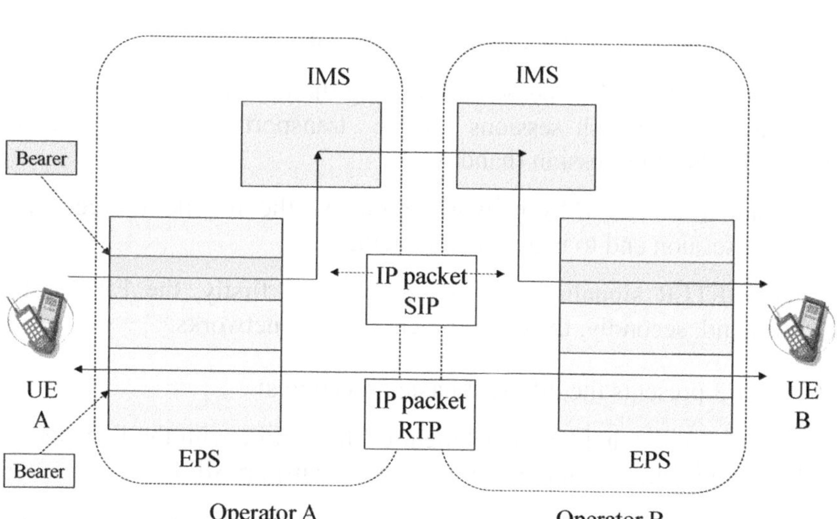

Figure 1. *Implementation of VoLTE or ViLTE services*

The EPS network does not provide telephone service because it does not deal with telephone signaling.

The EPS network operates in packet-switched (PS) mode and acts as the transport of internet protocol (IP) packets through bearers.

The EPS network, therefore, transfers the IP packets containing voice or video real-time transport protocol (RTP) streams or telephone signaling session initiation protocol (SIP).

Telephone or videophone service is provided by the IMS network which provides the functions as follows:

– routing the call;

– supplementary telephone and videophone services;

– interconnection to the third-party networks.

Chapter 1 presents the architecture of EPS and IMS networks and these networks environment: databases, charging, policy and charging control (PCC), DIAMETER routing, ENUM system and internet protocol exchange (IPX).

Chapter 2 presents various signaling protocols:

– signaling of the EPS network, allowing the mobile to attach, to update its location, to establish sessions for the transport of IP packets and to change cells during a session (handover);

– signaling of the IMS network, allowing the mobile to register, to establish a session and to negotiate the media;

– DIAMETER signaling exchanged between, firstly, the EPS or IMS networks, and, secondly, the environment of these networks.

Chapter 3 presents the different basic procedures:

– the attachment and the detachment of the mobile with the EPS network and the establishment of the default bearer to transport SIP flows;

– the registration and the deregistration of the mobile with the IMS network;

– the establishment and the release of VoLTE and ViLTE session.

Chapter 4 presents the characteristics of the radio interface, for which the following features are described: data structure, transmission chain of the physical layer, frequency time and space multiplexing.

The same chapter also illustrates two procedures of the radio interface: access control of the mobile to network and data transfer.

Chapter 5 presents the supplementary telephone and videophone services offered by a particular entity of the IMS network, the telephony application server (TAS).

These services include call forwarding, identity presentation, message waiting indication, call hold, conference call, call waiting and call barring.

It also presents the characteristics of audio and video streams.

Chapter 6 presents the interconnection to the public switched telephone network (PSTN) or to the public land mobile network (PLMN) (Figure 2).

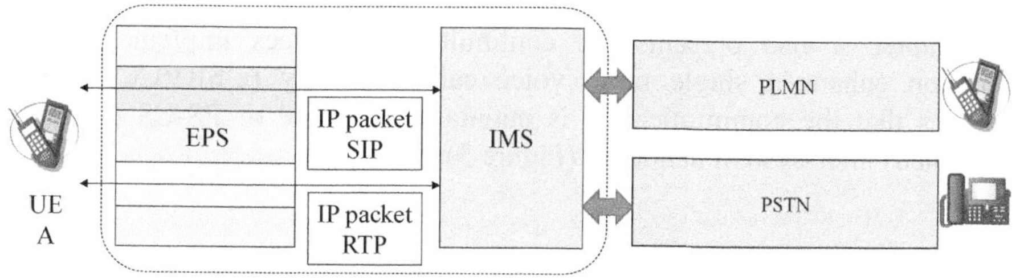

Figure 2. *Interconnection to the PSTN and PLMN network*

Chapter 6 also presents the interconnection of the IMS network with IMS third-party networks.

Chapter 7 presents the mechanisms of intra-system and PS-PS inter-system handover.

The intra-system handover is performed when the mobile changes cell but does not change the 4G network concerned.

The PS-PS inter-system handover is performed when the mobile changes cell and network but holds the PS mode. This type of handover is applied to VoLTE or ViLTE services if the same functionality exists in the HSPA evolution of 3G network.

Both handover modes are transparent to VoLTE and ViLTE services, the movement of the mobile being masked for the IMS network.

Chapter 8 presents the roaming for which two routing methods of the RTP streams are described:

– nominal routeing of the RTP stream that passes through the home network;

– optimal routeing of the RTP stream that does not pass through the home network.

Chapter 9 presents the centralization of services implemented by IMS centralized services (ICS) that enables the IMS network to offer VoLTE and ViLTE services regardless of the network where the mobile phone is connected.

Chapter 9 also presents the continuity of services implemented by function enhanced single radio voice call continuity (e-SRVCC) which ensures that the communication is maintained in case of PS-CS (Circuit-Switched) inter-system handover (Figure 3).

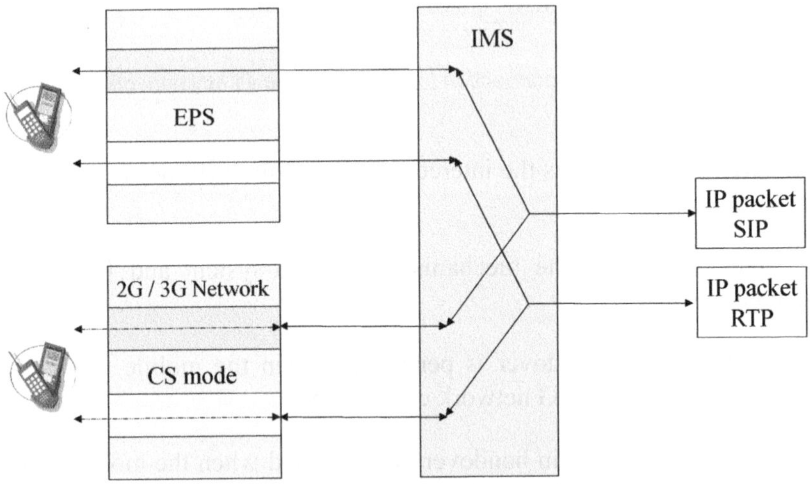

Figure 3. *PS-CS inter-system handover*

Chapter 10 presents the two modes providing short message service (SMS).

Short message service over SGsAP allows a mobile connected to the 4G network to send and receive SMS in the CS mode.

Short message service over SIP is a supplementary telephone service provided by the IMS network.

André PEREZ
April 2016

List of Abbreviations

A

AAA	Authorization-Authentication-Answer
AAR	Authorization-Authentication-Request
ACA	Accounting-Answer
ACM	Address Complete Message
ACR	Accounting-Request
AF	Application Function
AIA	Authentication-Information-Answer
AIR	Authentication-Information-Request
AM	Acknowledged Mode
AMBR	Aggregate Maximum Bit Rate
AMR	Adaptive Multi-Rate
AMR WB	AMR Wide Band
ANM	Answer Message
AOC	Advice of Charge
APM	Application transport Mechanism
APN	Access Point Name
ARP	Allocation and Retention Priority
ARQ	Automatic Repeat Request

AS	Application Server
ASA	Abort-Session-Answer
ASR	Abort-Session-Request
ATCF	Access Transfer Control Function
ATGW	Access Transfer Gateway
ATU-STI	Access Transfer Update – Session Transfer Identifier
AUTN	Authentication Network

B

B2BUA	Back-to-Back User Agent
BCCH	Broadcast Control Channel
BCH	Broadcast Channel
BCTP	Bearer Control Tunnelling Protocol
BGCF	Breakout Gateway Control Function
BICC	Bearer Independent Call Control
BSR	Buffer Status Report
BSS	Base Station Sub-system

C

CA	Carrier Aggregation
CAP	Camel Application Part
CAT	Customized Alerting Tone
CBP	Constrained Baseline Profile
CC	Component Carrier
CCA	Credit-Control-Answer
CCBS	Completion of Communications to Busy Subscriber
CCCH	Common Control Channel
CCNL	Completion of Communications on Not Logged-in

CCNR	Completion of Communications on No Reply
CCR	Credit-Control-Request
CD	Communication Deflection
CDF	Charging Data Function
CDIV	Communication Diversion
CDR	Charging Data Record
CFB	Communication Forwarding on Busy User
CFI	Control Format Indicator
CFNL	Communication Forwarding on Not Logged-in
CFNR	Communication Forwarding on no Reply
CFU	Communication Forwarding Unconditional
CGF	Charging Gateway Function
CK	Cipher Key
CLA	Cancel-Location-Answer
CLR	Cancel-Location-Request
CM	Call Management
CMAS	Commercial Mobile Alert System
CNG	Comfort Noise Generation
CP	Cyclic Prefix
CQI	Channel Quality Indicator
CRI	Contention Resolution Identity
C-RNTI	Cell RNTI
CRS	Customised Ringing Signal
CS	Circuit-Switched
CSCF	Call Session Control Function
CSFB	CS FallBack
CTF	Charging Trigger Function
CUG	Closed User Group
CW	Communication Waiting

D

DCCH	Dedicated Control Channel
DCI	Downlink Control Information
DDA	Delete-Subscriber-Data-Answer
DDR	Delete-Subscriber-Data-Request
DEA	DIAMETER Edge Agent
DL-SCH	Downlink Shared Channel
DNS	Domain Name System
DRB	Data Radio Bearer
DM-RS	Demodulation Reference Signal
DRA	DIAMETER Routing Agent
DRX	Discontinuous Reception
DSCP	DiffServ Code Point
DTCH	Dedicated Traffic Channel
DTX	Discontinuous Transmission
DwPTS	Downlink Pilot Time Slot

E

EATF	Emergency Access Transfer Function
ECGI	E-UTRAN Cell Global Identifier
E-CSCF	Emergency-CSCF
ECT	Explicit Communication Transfer
EM	End Marker
EMM	EPS Mobility Management
eNB	evolved Node B
EPC	Evolved Packet Core
EPS	Evolved Packet System
E-RAB	EPS Radio Access Bearer

ESM	EPS Session Management
e-SRVCC	enhanced Single Radio Voice Call Continuity
ETWS	Earthquake and Tsunami Warning System
E-UTRAN	Evolved Universal Terrestrial Radio Access Network
EVS	Enhanced Voice Services

F

FA	Flexible Alerting
FB	Full Band
FDD	Frequency Division Duplex
FFT	Fast Fourier Transform
FR	Full Rate

G

GBR	Guaranteed Bit Rate
GGSN	Gateway GPRS Support Node
GMSC	Gateway MSC
GP	Gap Period
GPRS	General Packet Radio Service
GSM	Global System for Mobile
GTP-C	GPRS Tunnel Protocol Control
GTP-U	GPRS Tunnel Protocol User
GUTI	Globally Unique Temporary Identity

H

HARQ	Hybrid ARQ
HI	HARQ Indicator
HII	High Interference Indication
HLR	Home Location Register

H-PCRF	Home PCRF
HR	Half Rate
HSS	Home Subscriber Server
HTTP	Hypertext Transfer Protocol

I

IAM	Initial Address Message
IBCF	Interconnection Border Control Function
ICB	Incoming Communication Barring
ICS	IMS Centralized Services
ICIC	Inter-Cell Interference Coordination
I-CSCF	Interrogating-CSCF
IDA	Insert-Subscriber-Data-Answer
IDR	Insert-Subscriber-Data-Request
IETF	Internet Engineering Task Force
iFC	initial Filter Criteria
IFFT	Inverse Fast Fourier Transform
IK	Integrity Key
IMPI	IMS Private User Identity
IMPU	IMS Public User Identity
IMRN	IP Multimedia Routing Number
IMS	IP Multimedia Sub-system
IMS-GWF	IMS Gateway Function
IMSI	International Mobile Subscriber Identity
IOI	Interference Overload Indication
IP	Internet Protocol
IPBCP	IP Bearer Control Protocol
IPSec	IP Security
IP-SM-GW	IP Short Message Gateway

IPX	Internet Protocol eXchange
ISC	IMS Service Control
ISIM	IMS Services Identity Module
ISUP	ISDN User Part
IWMSC	Inter Working MSC

L

LAI	Location Area Identifier
LCID	Logical Channel Identifier
LIA	Location-Info-Answer
LIR	Location-Info-Request
LRF	Location Retrieval Function
LTE	Long Term Evolution

M

MAA	Multimedia-Auth-Answer
MAC	Media Access Control
MAR	Multimedia-Auth-Request
MBR	Maximum Bit Rate
MBSFN RS	MBMS Single Frequency Network RS
MCC	Mobile Country Code
MCCH	Multicast Control Channel
MCH	Multicast Channel
MCID	Malicious Communication Identification
MGCF	Media Gateway Control Function
MGW	Multimedia Gateway
MIB	Master Information Block
MIMO	Multiple Input Multiple Output
MISO	Multiple Input Single Output

MME	Mobility Management Entity
MNC	Mobile Network Code
MP	Main Profile
MRF	Multimedia Resource Function
MRFC	MRF Controller
MFRP	MRF Processor
MSC	Mobile-services Switching Centre
MDISDN	Mobile Subscriber ISDN Number
MTCH	Multicast Traffic Channel
MWI	Message Waiting Indication

N

NAPT	Network Address and Port Translation
NAPT-PT	NAPT Protocol Translation
NAS	Non Access Stratum
NB	Narrow Band
NOA	Notify-Answer
NOR	Notify-Request

O

OCB	Outgoing Communication Barring
OCS	Online Charging System
OFCS	Offline Charging System
OFDM	Orthogonal Frequency-Division Multiplexing
OFDMA	Orthogonal Frequency-Division Multiple Access
OIP	Originating Identification Presentation
OIR	Originating Identification Restriction
OMR	Optimal Media Routeing
OTDOA	Observed Time Difference of Arrival

P

PBCH	Physical Broadcast Channel
PCC	Policy and Charging Control
PCCH	Paging Control Channel
PCEF	Policy and Charging Enforcement Function
PCFICH	Physical Control Format Indicator Channel
PCH	Paging Channel
PCI	Physical-layer Cell Identity
PCRF	Policy Charging and Rules Function
P-CSCF	Proxy-CSCF
PDCCH	Physical Downlink Control Channel
PDCP	Packet Data Convergence Protocol
PDN	Packet Data Network
PDSCH	Physical Downlink Shared Channel
PGW	PDN Gateway
PHICH	Physical HARQ Indicator Channel
PHR	Power Headroom Report
PLMN	Public Land Mobile Network
PMCH	Physical Multicast Channel
PMI	Precoding Matrix Indicator
PNA	Push-Notification-Answer
PNR	Push-Notification-Request
PPA	Push-Profile-Answer
PPR	Push-Profile-Request
PRACH	Physical Random Access Channel
PRS	Positioning Reference Signal
PS	Packet-Switched
PSAP	Public Safety Answering Point
PSI	Public Service Identity

PSS	Primary Synchronization Signal
PSTN	Public Switched Telephone Network
PUCCH	Physical Uplink Control Channel
PUA	Profile-Update-Answer
PUR	Profile-Update-Request
PUSCH	Physical Uplink Shared Channel

Q

QAM	Quadrature Amplitude Modulation
QCI	QoS Class Identifier
QoS	Quality of Service
QPSK	Quadrature Phase-Shift Keying

R

RAA	Re-Auth-Answer
RACH	Random Access Channel
RAR	Random Access Response
RAR	Re-Auth-Request
RA-RNTI	Random Access RNTI
RAT	Radio Access Technology
RB	Resource Block
RE	Resource Element
REL	Release
RFC	Request For Comments
RI	Rank Indicator
RLC	Radio Link Control
RLC	Release Complete
RNC	Radio Network Controller
RNTI	Radio Network Temporary Identity

RNTP	Relative Narrowband Tx Power
ROHC	Robust Header Compression
RRC	Radio Resource Control
RS	Reference Signal
RSA	Reset-Answer
RSR	Reset-Request
RSRP	Reference Signal Received Power
RSRQ	Reference Signal Received Quality
RTA	Registration-Termination-Answer
RTP	Real-time Transport Protocol
RTR	Registration-Termination-Request
RV	Redundancy Version

S

SAA	Server-Assignment-Answer
SAR	Server-Assignment-Request
SCC AS	Service Centralization and Continuity AS
SC-FDMA	Single Carrier Frequency Division Multiple Access
S-CSCF	Serving-CSCF
SDF	Service Data Flow
SDP	Session Description Protocol
SGSN	Service GPRS Support Node
SFN	System Frame Number
SGW	Serving Gateway
SIB	System Information Block
SIGTRAN	Signalling Transport over IP
SIMO	Single Input Multiple Output
SIP	Session Initiation Protocol
SIP-I	SIP with Encapsulated ISUP

SI-RNTI	System Information RNTI
SISO	Single Input Single Output
SLF	Subscription Locator Functional
SM-AL	Short Message Application Layer
SM-CL	Short Message Control Layer
SM-RL	Short Message Relay Layer
SM-TL	Short Message Transport Layer
SMS	Short Message Service
SMS-SC	SMS Service Center
SNA	Subscribe-Notifications-Answer
SNR	Subscribe-Notifications-Request
SPR	Subscription Profile Repository
SPS	Semi-Persistent Scheduling
SRB	Signalling Radio Bearer
SRS	Sounding Reference Signal
SS7	Signalling System 7
SSS	Secondary Synchronization Signal
STA	Session-Termination-Answer
S-TMSI	Shortened-TMSI
STN-SR	Session Transfer Number for SRVCC
STR	Session-Termination-Request
SWB	Super Wide Band

T

TA	Timing Advance
TAI	Tracking Area Identity
TAS	Telephony Application Server
TC-RNTI	Temporary Cell RNTI
TDD	Time Division Duplex

TDM	Time Division Multiplexing
TEID	Tunnel Endpoint Identifier
THIG	Topology Hiding Interconnect Gateway
TIP	Terminating Identification Presentation
TIR	Terminating Identification Restriction
TM	Transparent Mode
TMSI	Temporary Mobile Subscriber Identity
TPC	Transmit Power Control
TRF	Transit and Roaming Function
TrGW	Transition Gateway
TTI	Transmission Time Interval

U

UA	User Agent
UAA	User-Authorization-Answer
UAC	User Agent Client
UAR	User-Authorization-Request
UAS	User Agent Server
UCI	Uplink Control Information
UDA	User-Data-Answer
UDR	User-Data-Request
UE	User Equipment
UICC	Universal Integrated Circuit Card
ULA	Update-Location-Answer
ULR	Update-Location-Request
UL-SCH	Uplink Shared Channel
UM	Unacknowledged Mode
UMTS	Universal Mobile Telecommunications System
UpPTS	Uplink Pilot Time Slot

URI	Uniform Resource Identifier
URN	Uniform Resource Name
USIM	Universal Services Identity Module
UTRAN	Universal Terrestrial Radio Access Network

V

VAD	Voice Activity Detection
ViLTE	Video over LTE
VoHSPA	Voice over High Speed Packet Access
VoLTE	Voice over LTE
V-PCRF	Visited PCRF

W, X

WB	Wide Band
XCAP	XML Configuration Access Protocol
XML	eXtensible Markup Language

1

Network Architecture

1.1. EPS network

1.1.1. *Functional architecture*

The functional architecture of the evolved packet system (EPS) network is illustrated in Figure 1.1.

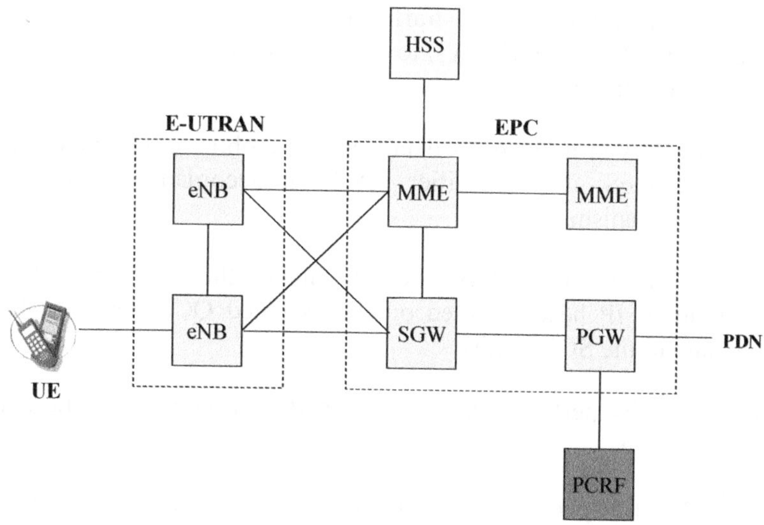

Figure 1.1. *Functional architecture of EPS network*

The EPS mobile network consists of an evolved packet core (EPC) network and an evolved universal terrestrial radio access network (E-UTRAN).

The E-UTRAN access network ensures the connection of the User Equipment (UE).

The EPC core network interconnects the access networks, provides the interface to the packet data network (PDN) and ensures the attachment of mobile phones and the establishment of bearers.

1.1.1.1. *eNB entity*

The E-UTRAN access network includes a single type of entity, the evolved Node Base station (eNB) that connects to the mobiles.

The eNB entity is responsible for the management of radio resources, for the control of the establishment of the data radio dearer (DRB), in which the mobile traffic is transmitted and for its mobility management during the session (handover).

The eNB entity transfers the traffic data from the mobile (respectively from the Serving Gateway (SGW)) to the SGW entity (to the mobile phones concerned, accordingly).

When the eNB entity receives data from the mobile or the SGW entity, it refers to the QoS class identifier (QCI) in accordance with the data scheduling mechanism.

The eNB entity can perform the marking of the DiffServ code point (DSCP) field of IP header, based on the assigned QCI identifier, for the outgoing data to the SGW entity.

The eNB entity performs compression and encryption of traffic data on the radio interface.

The eNB entity performs encryption and integrity control of signaling data exchanged with the mobile.

It also undertakes the selection of the mobility management entity (MME) to which the mobile is attached.

It treats paging requests sent by the MME entity for their distribution in the cellphone corresponding to the radio coverage area of the eNB entity.

The eNB entity also distributes system information to the cell containing the technical characteristics of the radio interface and allowing the mobile access to connect.

The eNB entity uses the measurements made by the mobile to decide on the initiation of a cell change during a session (handover).

1.1.1.2. *MME entity*

The MME entity is the network control tower, allowing mobile access and controlling bearer establishment for the transmission of traffic data.

The MME entities belong to a group (pool). Load balancing of MME entities is provided by the eNB entities within a group that must have access to each MME entity of the same group.

The MME entity is responsible for attachment and detachment of the mobile phone to the network concerned.

During attachment, the MME entity retrieves the subscriber's profile and the subscriber's authentication data stored in the home subscriber server (HSS) and performs authentication of the mobile.

During attachment, the MME entity registers the tracking area identity (TAI) of the mobile and allocates a globally unique temporary identity (GUTI) to the mobile which replaces the private international mobile subscriber identity (IMSI).

The MME entity manages a list of location areas allocated to the mobile, where the mobile can move in an idle state, without contacting the MME entity to update its TAI location area.

When attaching the mobile, the MME selects SGW and PGW (PDN Gateway) entities for the construction of the default bearer, e.g. for the transport of IP packets containing Session Initiation Protocol (SIP) signaling.

For the construction of the bearer, the selection of the PGW entity is obtained from the access point name (APN), communicated by the mobile or by the HSS entity in the subscriber's profile.

The source MME entity also selects the target MME entity when the mobile changes both cell and group (pool).

The MME entity provides the information required for lawful interception, such as the mobile status (idle or connected), the TAI location area if the mobile is idle or the E-UTRAN Cell Global Identifier (ECGI) if the mobile is in session.

1.1.1.3. *SGW entity*

The SGW entities are organized into groups (pools). To ensure load balancing of SGW entities, each eNB entity within a group must have access to each SGW entity of the same group.

The SGW entity forwards incoming data from the PGW entity to the eNB entity and outgoing data from the eNB entity to the PGW entity.

When the SGW entity receives data from the eNB or PGW entities, it refers to the QCI identifier for the implementation of the data scheduling mechanism.

The SGW entity can perform marking of the DSCP field of IP header based on the assigned QCI identifier for incoming and outgoing data.

The SGW entity is the anchor point for intra-system handover (mobility within EPS network) provided that the mobile does not change group. Otherwise, the PGW entity performs this function.

The SGW entity is also the anchor point at the inter-system handover PS-PS, requiring the transfer of traffic data from the mobile to the second or third generation mobile network.

The SGW entity informs the MME entity of incoming data when the mobile is in idle state, which allows the MME entity to trigger paging of all eNB entities of the TAI location area.

A mobile in the idle state remains attached to the MME entity. However, it is no longer connected to the eNB entity and thus the radio bearer and the S1 bearer are deactivated.

1.1.1.4. *PGW entity*

The PGW entity is the gateway router providing the EPS network connection to the PDN network.

When the PGW entity receives data from the SGW entity or PDN network, it refers to the QCI identifier for the implementation of the data scheduling mechanism.

The PGW entity can perform DSCP marking of IP header based on the assigned QCI identifier.

During attachment, the PGW entity grants an IPv4 or IPv6 address to the mobile.

The PGW entity constitutes the anchor point for inter-SGW mobility when the mobile changes groups.

The PGW entity hosts the policy and charging enforcement function (PCEF) which applies the rules relating to mobile traffic data on packet filtering, on charging and on quality of service (QoS) to be applied to the bearer to build.

The policy charging and rules function (PCRF) entity, outside the EPS network, provides the PCEF function of the PGW entity with the rules to apply when establishing bearers.

The PGW entity generates data for use by charging entities to develop the record tickets processed through the billing system.

The PGW entity performs replication of the mobile traffic data within the framework of lawful interception.

1.1.2. *Protocol architecture*

The protocol architecture of the EPS network is illustrated in Figure 1.2 for the control plane and in Figure 1.3 for the traffic plane.

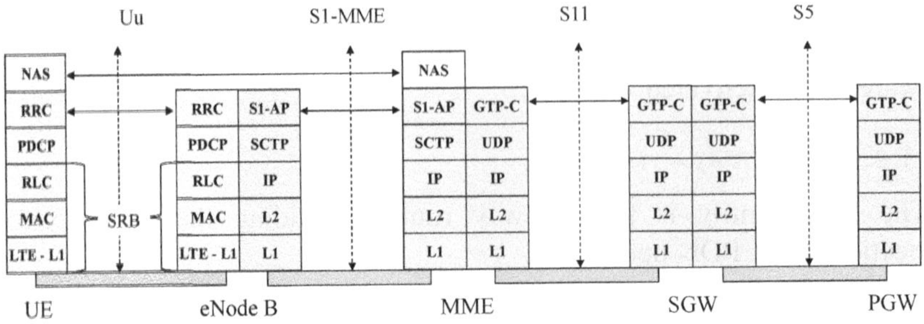

Figure 1.2. *Protocol architecture: control plane*

The LTE-Uu interface is the reference point between the mobile and the eNB entity.

This interface supports radio resource control (RRC) signaling exchanged between the mobile and the eNB entity, transmitted in the signaling radio bearer (SRB) and the mobile traffic data transmitted in the data radio bearer (DRB).

The RRC signaling also provides transport of the non-access stratum (NAS) protocol exchanged between the mobile and the MME entity.

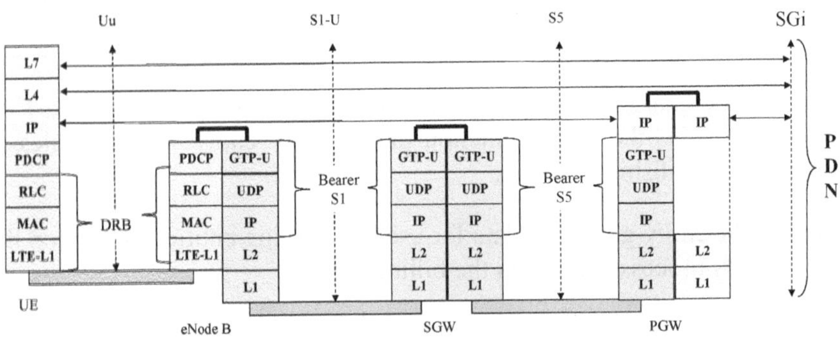

Figure 1.3. *Protocol architecture: traffic plane*

The S1-MME interface is the reference point between the MME and eNB entities for signaling, via the S1-AP (Application Part) protocol.

The S1-AP protocol also provides transport of the NAS protocol exchanged between the mobile and the MME entity.

The S11 interface is the reference point between the MME and SGW entities for signaling via the GPRS (General Packet Radio Service) tunnel control protocol (GTPv2-C).

The S5 interface is the reference point between the SGW and PGW entities for signaling via the GTPv2-C protocol and traffic data (IP packets) via the GPRS tunnel protocol user (GTP-U).

The S10 interface is the reference point between the MME entities for signaling, via the GTPv2-C protocol.

The S1-U interface is the reference point between the eNB and SGW entities for traffic data, via the GTP-U protocol.

The SGi interface is the reference point between the PDW entity and the PDN network (Internet).

The X2 interface is the reference point between two eNB entities for signaling, via the X2-AP protocol (Figure 1.4) and for incoming traffic data via the GTP-U protocol when mobile changes cells (Figure 1.5).

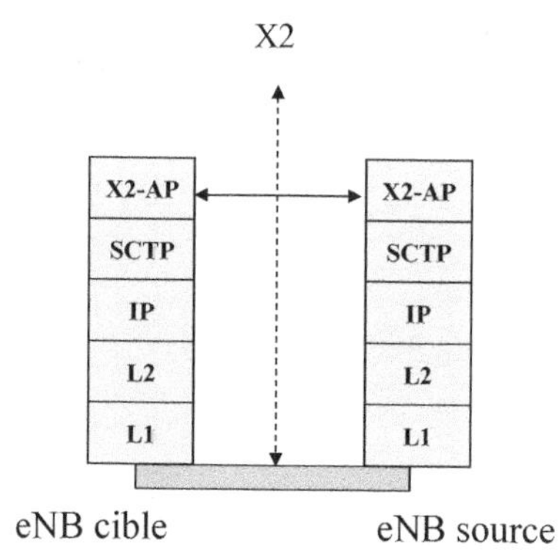

Figure 1.4. *Protocol architecture of the X2 interface: control plane*

The bearer established between the two eNB entities is unidirectional (eNB source to eNB target). It allows for the transfer of traffic data received from the SGW entity to the target eNB entity. It is established temporarily, for the time of the handover of the mobile.

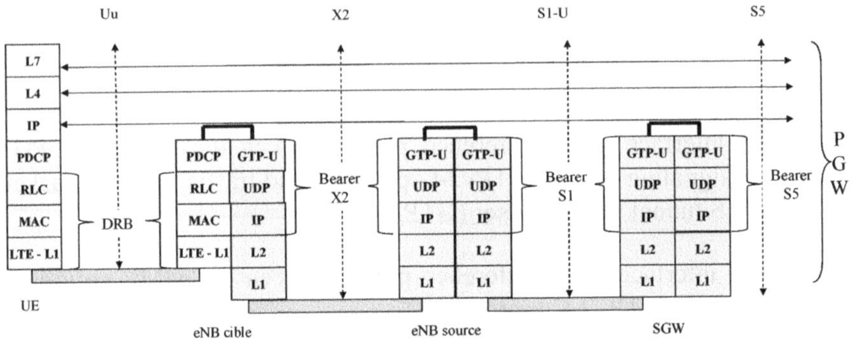

Figure 1.5. *Protocol architecture traffic plane during handover based on the X2 interface*

1.1.3. *Bearers*

1.1.3.1. *Bearer structure*

The EPS network carries traffic data (IP packets) transparently to the PGW entity that performs packet routing.

The IP packet is carried in bearers constructed between EPS network entities (Figure 1.6).

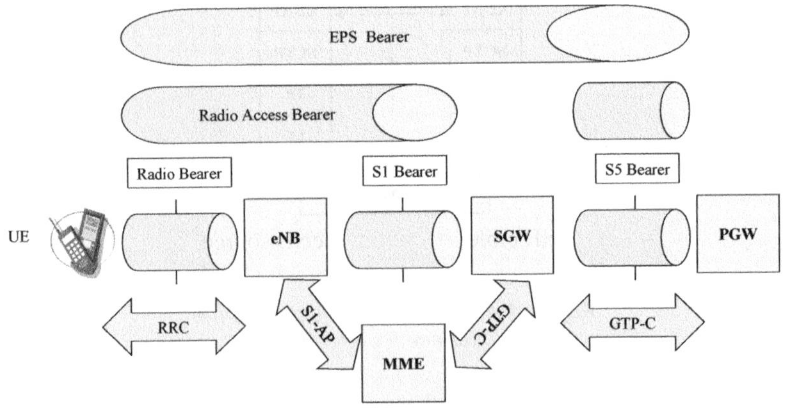

Figure 1.6. *Construction of the bearers*

The data radio bearer (DRB) is constructed between the user equipment (UE) and the eNB entity. The RRC signaling, which is exchanged between the mobile and the eNB entity, is responsible for constructing this bearer.

The S1 bearer is constructed between the eNB and SGW entities. The S1-AP signaling, exchanged between the eNB and MME entities and the GTPv2-C signaling, exchanged between the MME and SGW entities, are responsible for constructing this bearer.

The S5 bearer is constructed between the SGW and PGW entities. The GTPv2-C signaling exchanged between the SGW and PGW entities is responsible for constructing this bearer.

The connection of the radio bearer and the S1 bearer, performed by the eNB entity, constitutes the EPS radio access bearer (E-RAB).

The connection of the E-RAB and S5 bearers, which is performed by the SGW entity, constitutes the EPS bearer.

The PGW entity is the only EPS mobile network entity that routes traffic data (IP packets).

The IP transport network that enables communication between the EPS network entities routes the S1 or S5 bearers.

The eNB and SGW entities do not perform routeing. They only provide the connection between the bearers.

There are two types of bearers in the EPS network:

– the default bearer established when attaching the mobile, used for example to transport SIP signaling;

– the dedicated bearer, established following a specific request from the mobile, used for example for transport of real time protocol (RTP) streams containing voice or video.

1.1.3.2. *Quality of Service*

The EPS bearer can be of guaranteed bit rate (GBR) type or can be of non-GBR type.

Table 1.1 provides the QoS characteristics associated with these two bearer families.

QoS characteristics	*GBR*	*Non-GBR*
QCI (QoS Class Identifier)	✓	✓
ARP (Allocation and Retention Priority)	✓	✓
GBR (Guaranteed Bit Rate)	✓	
MBR (Maximum Bit Rate)	✓	
APN-AMBR (Aggregate Maximum Bit Rate)		✓
UE-AMBR		✓

Table 1.1. *QOS characteristics*

The QCI parameter indicates the priority level, the delay and the packet loss rate (Table 1.2).

QCI from 1 to 4 are assigned to GBR bearers. QCI from 5 to 9 are assigned to non-GBR bearers.

QCI	*Resource Type*	*Priority*	*Delay*	*Packet loss rate*	*Examples of services*
1	GBR	2	100 ms	10^{-2}	Voice
2		4	150 ms	10^{-3}	Video Calling
3		3	50 ms	10^{-3}	Games
4		5	300 ms	10^{-6}	Video.
5	Non-GBR	1	100 ms	10^{-6}	SIP Signaling
6		6	300 ms	10^{-6}	Video, Internet
7		7	100 ms	10^{-3}	Voice, Video, Games
8		8	300 ms	10^{-6}	Video, Internet
9		9			

Table 1.2. *QCI parameters*

The scheduling of traffic data carried out at the level of the eNB, SGW and PGW entities is based on the QCI priority level.

The bit rate control is done from the GBR and MBR parameters for guaranteed bit rate bearers.

The bit rate control is conducted for each bearer at the eNB and PGW entities for incoming data in the EPS network.

The bit rate control is done from the APN-AMBR and UE-AMBR parameters for non-GBR type bearers. This control is performed for aggregated bit rates of non-GBR bearers of a mobile.

The APN-AMBR parameter controlled by the PGW entity corresponds to the maximum bit rate authorized for all the streams of a mobile phone using non-GBR bearers at the PGW entity level.

The UE-AMBR parameter controlled by the eNB entity corresponds to the maximum authorized bit rate for all streams of a mobile phone using non-GBR bearers, at the eNB entity level.

The pre-emption implemented at the eNB and PGW entity level corresponds to the ARP parameter that defines the following information:

– pre-emption capability: this parameter is used for the establishment of a new session, if the resource is not available. This parameter determines whether or not a new session can pre-empt an existing session;

– pre-emption vulnerability: this parameter is used by the existing session. This parameter is compared to the pre-emption capability parameter of the new session to determine whether the existing session can be pre-empted or not;

– priority: this determines the priority level associated with pre-emption. This priority level is independent of that set for the QCI parameter.

The QoS parameters (QCI, ARP and APN-AMBR) relating to default bearers are stored in the HSS entity. These values can be changed by the PCRF entity.

The QoS parameters (QCI, ARP, GBR and MBR) relating to dedicated bearers are stored in the subscription profile repository (SPR) entity associated with the PCRF entity.

The MME entity replaces the UE-AMBR parameter provided by the HSS entity by the sum of the different APN-AMBR parameters, provided it is less than the value indicated by the HSS entity.

1.2. IMS network

1.2.1. *Functional architecture*

The functional architecture of the IP multimedia subsystem (IMS) network is described in Figure 1.7.

The IMS network includes the following entities:

– call session control function (CSCF), involving P-CSCF (Proxy-CSCF), S-CSCF (Serving-CSCF), I-CSCF (Interrogating-CSCF) and E-CSCF (Emergency-CSCF);

– application servers (AS), involving telephony application server (TAS);

– multimedia resource function (MRF), involving MRFC (MRF Controller) and MFRP (MRF Processor).

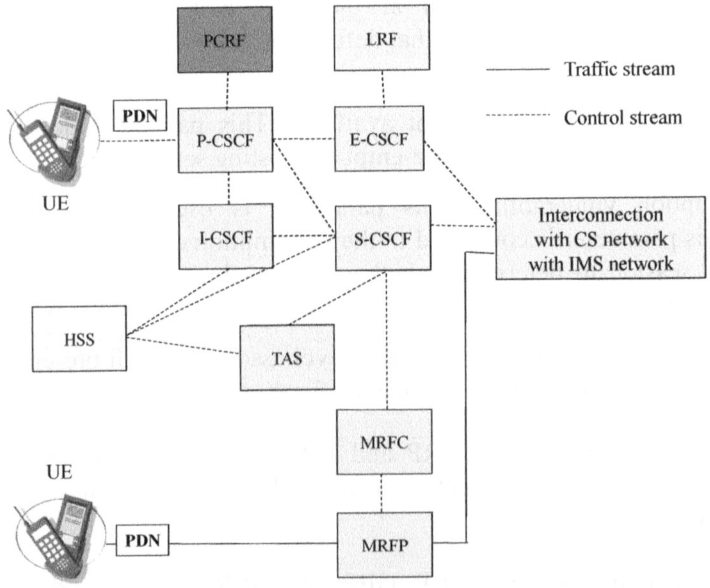

Figure 1.7. *Functional architecture of IMS network*

1.2.1.1. *P-CSCF entity*

The P-CSCF entity is the first point of contact for the mobile in the IMS network. It performs the function of a PROXY SERVER. It receives the requests from the UE or from the S-CSCF entity and transfers them respectively to the S/I-CSCF entity or to the UE entity.

The P-CSCF entity may also act as an user agent (UA) in abnormal operating conditions, when it has to terminate or generate SIP transactions.

During registration, the P-CSCF entity forwards the SIP REGISTER request to the I-CSCF entity determined on the basis of the domain name provided by the UE entity. Into this message, it adds a header *path* containing its identity. This identity is preserved by the S-CSCF entity.

During session establishment, the P-CSCF entity forwards the SIP INVITE request received from the S-CSCF entity (or from the UE entity respectively) to the UE entity (or respectively to the S-CSCF entity).

To carry out the transfer, the P-CSCF entity has to retrieve the IP addresses of the UE entity (or respectively of the S-CSCF entity):

– the SIP INVITE request received from the S-CSCF entity contains the IP address of the UE entity instead of the uniform resource identifier (URI);

– the SIP INVITE request received from the UE entity contains the identity of the S-CSCF entity in the header *route*.

The P-CSCF entity detects emergency calls and forwards them to the E-CSCF entity.

The P-CSCF entity generates the data necessary for the generation of charging tickets.

The P-CSCF entity establishes an IPSec (IP Security) association with the UE at its registration.

During session establishment, the P-CSCF entity controls the type of resources required by the UE on the basis of the capacities authorized by the EPS network, using the DIAMETER messages exchanged with the PCRF entity.

During session establishment, the P-CSCF entity checks resource availability in the EPS network.

1.2.1.2. *I-CSCF entity*

The I-CSCF entity is the specific point of contact within the IMS network for some transactions coming from the P-CSCF or S-CSCF entities. It performs the function of a PROXY SERVER.

Upon receipt of the first SIP REGISTER request, the I-CSCF assigns an S-CSCF entity to the UE entity and transfers the request to the S-CSCF entity chosen. To fulfill this function, an exchange of DIAMETER messages with the HSS entity is necessary.

Upon receipt of the second SIP REGISTER request and the first SIP INVITE request, for an incoming call, the I-CSCF entity queries the HSS entity for the IP address of the S-CSCF entity attributed to the UE entity and transfers the request to that S-CSCF entity.

The I-CSCF entity forwards the data necessary for the generation of charging tickets.

1.2.1.3. *S-CSCF entity*

The S-CSCF entity provides the UE entity with session control services. It performs different roles depending on the type of request received:

– that of a REGISTRAR for the registration of the UE entity;

– that of a LOCATION SERVER for the storage of the correspondence between the IP address and the URI identity of the UE entity;

– that of a PROXY SERVER for the establishment of a session;

– that of an UA, in abnormal operating conditions, when it has to terminate or generate SIP transactions.

On receiving the first REGISTER request, the S-CSCF entity contacts the HSS entity to recover the UE authentication data. To fulfill this function, an exchange of DIAMETER messages with the HSS entity is required. The S-CSCF entity responds with a message *401 unauthorized* containing the parameters used for authentication.

On receiving the second REGISTER request, The S-CSCF entity authenticates the UE and recovers its profile from the HSS entity. It responds with a message *200 OK* containing a header *Service Route* with its IP address which the UA keeps in its memory.

For an outgoing call, on receipt of the first INVITE request from the P-CSCF entity, the S-CSCF entity performs a check on the service required on the basis of the profile recovered during registration. The S-CSCF entity transfers the request either to the application server or to the entity allocated to the interconnection. The IP address of the application server is contained in the profile of the UE entity recovered during registration.

For an incoming call, on receipt of the first INVITE request from the I-CSCF entity, the S-CSCF entity performs a check on the service required. The S-CSCF entity transfers the request either to the application server or to the P-CSCF entity.In S-CSCF, as suggested in the latter case, the entity replaces the URI identity of the request with the IP address of the UE entity. The IP address of the P-CSCF entity is recovered on the basis of the header *Path*, during the registration of the UE entity.

The S-CSCF entity forwards the data necessary to generate charging tickets.

1.2.1.4. *E-CSCF entity*

The E-CSCF entity handles emergency calls transmitted by the P-CSCF and routes the request to the public-safety answering point (PSAP) nearest to the UE. The PSAP emergency center can be linked to a fixed or mobile network or to another IMS network.

Upon receiving the INVITE request, the E-CSCF entity retrieves the location of the mobile in the header *P-Access-Network-Info*.

The header *P-Access-Network-Info* was inserted by the P-CSCF entity. The value of the mobile location was provided by the PGW entity through the PCRF entity.

On receiving the INVITE request, the E-CSCF entity contacts the location retrieval function (LRF) to obtain the URI identity of the PSAP emergency center.

On the basis of information provided by the LRF entity, the E-CSCF entity transfers the request to the entity allocated to the interconnection to the CS network or the IMS network.

1.2.1.5. *Application servers*

The application server provides added-value services to the IMS network. For instance, it hosts and executes the supplementary telephone services. The AS entity may affect the SIP session depending on the service required.

The S-CSCF entity has to decide whether an application server is necessary for a specific treatment of an SIP request to ensure handling of the appropriate service. The decision is based on the information received from the HSS during mobile registration.

The application server may play various roles in the processing of an SIP message:

– that of a PROXY SERVER. In this mode, the SIP request from the S-CSCF entity is sent to the application server. The application server can add, remove or modify headers in the SIP message;

– that of an User Agent Server (UAS) or a REDIRECT SERVER. In this mode, the response of the application server to the SIP request from the S-CSCF entity is `2xx`, `4xx`, `5xx` or `6xx` (UAS) or `3xx` (REDIRECT SERVER);

– that of an user agent client (UAC). In this mode, the application server generates the SIP request and transmits it to the S-CSCF entity.

– that of a Back to Back User Agent (B2BUA). In this mode, the application server receiving an SIP request from the S-CSCF entity terminates the dialog with the UAC entity at the originating side, and generates a new request with an UAS entity at the terminating side.

1.2.1.6. *Media processing*

Media processing is done by the MRF function, divided into two entities, the MRFC and MRFP.

The MRFC entity interprets information from the S-CSCF and controls the MRFP on the basis of this interpretation;

The MRFC entity forwards the data necessary for the generation of charging tickets.

The MFRP entity generates media flows under the control of the MRFC, such as telephone announcements.

The MFRP entity combines media flows to provide a conferencing service.

The MFRP entity can also perform particular treatments of the media flows, such as the transcoding of the audio signal.

1.2.2. *Protocol architecture*

The Gm interface is the reference point between the EU and P-CSCF entities. This interface supports SIP and Session Description Protocol (SDP) signaling.

The Ut interface is the reference point between the EU and TAS entities. This interface supports XML Configuration Access Protocol (XCAP) signaling, allowing the configuration of supplementary services by the mobile.

The Mw interface is the reference point between the CSCF entities. This interface supports SIP and SDP signaling.

The Mx interface is the reference point between, on the one hand, the CSCF entities and, the other hand, the interconnection with the Circuit-Switched (CS) network or the IMS network. This interface supports SIP and SDP signaling.

The Mr interface is the reference point between the S-CSCF and MRFC entities. This interface supports SIP and SDP signaling.

The Mb interface is the reference point between the UE and MRFP entities. This interface supports the RTP stream.

The IMS service control (ISC) interface is the reference point between the S-CSCF and TAS entities. This interface supports SIP and SDP signaling.

1.3. Databases

1.3.1. *Functional architecture*

The HSS entity is a database storing the data specific to each user. The main data stored include the identities of the users, the authentication parameters and the service profile.

The subscriber has a private identity IMSI used when attaching to the EPS network.

The subscriber has an IMS private identity (IMPI) used while registering to the IMS network, and an IMS public identity (IMPU) when establishing a voice or a conversational video call.

The authentication parameters are used to control access to the mobile for attachment to the EPS network or registration to the IMS network.

The service profile determines the services that mobile has subscribed.

The S-CSCF entity accesses the HSS entity to recover the authentication data and the service profile.

The I-CSCF entity accesses the HSS entity to retrieve the identity of the S-CSCF entity that has attached the mobile.

The TAS entity can access the HSS entity to retrieve service data necessary for the performance of the supplementary telephone service.

The subscription locator function (SLF) is a database that records the identity of the HSS entity where the mobile subscription was recorded, in the case where several HSS entities are deployed.

1.3.2. *Protocol architecture*

The S6a interface is the reference point between the MME and HSS entities. This interface supports the DIAMETER signaling.

The Cx interface is the reference point between, on the one hand, the I-CSCF or S-CSCF entities, while the HSS entity on the other. This interface supports the DIAMETER signaling.

The Sh interface is the reference point between the TAS and HSS entities. This interface supports the DIAMETER signaling.

The Dx interface is the reference point between, on the one hand, the I-CSCF or S-CSCF entities and the other hand, the SLF entity. This interface supports the DIAMETER signaling.

The Dh interface is the reference point between the SLF and TAS entities. This interface supports the DIAMETER signaling.

1.4. Charging associated with IMS network

1.4.1. *Functional architecture*

In the case of online charging, the user's account is consulted before granting the permission to use the network resource. That account is decreased during the communication. When it reaches zero, the communication is cut off. This mode of charging corresponds to pre-paid service.

1.4.1.1. *Offline charging*

The functional architecture of the offline charging system (OFCS) is described in Figure 1.8.

The charging trigger function (CTF) generates charging events based on the observation of the use of network resources. It is integrated in all the entities of the IMS network.

The charging data function (CDF) receives the charging data from the CTF function. The CDF function then uses these data to generate charging data records (CDR).

The charging data records produced by the CDF function are kept in the charging gateway function (CGF), a database which acts as a gateway with the billing system.

1.4.1.2. *Online charging*

The functional architecture of the online charging system (OCS) is described in Figure 1.9.

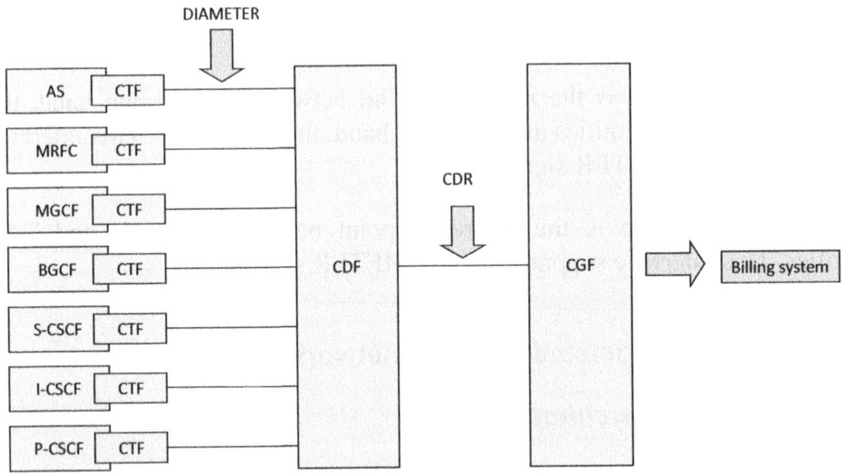

Figure 1.8. *Functional architecture of OFCS*

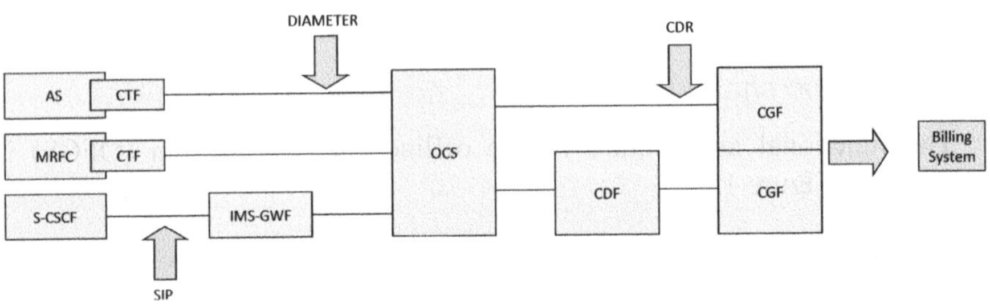

Figure 1.9. *Functional architecture of OCS*

The S-CSCF entity does not trigger any charging event and does not necessarily include the CTF function. Charging during a session is a service logic controlled by an application server IMS Gateway Function (IMS-GWF).

The OCS entity comprises several distinct modules:

– charging on the basis of the sessions established by the users (e.g. voice calls);

– charging on the basis of events in conjunction with application servers;

– valorization of the use of network resources to calculate the amount of charging;

– balance of the user's account.

The generation of CDRs sent to the billing system is optional. If the OCS entity does not produce CDRs, they are used by the same CDF as with offline charging.

1.4.2. *Protocol architecture*

The Rf interface is the reference point between, on the one hand, the entities of the IMS network, while on the other hand, the OFCS entity. This interface supports the DIAMETER signaling.

The Ro interface is the reference point between, on the one hand, the entities of the IMS network, and on the other hand, the OCS entity. This interface supports the DIAMETER signaling.

1.5. PCC function

1.5.1. *Functional architecture*

The functional architecture of the policy and charging control (PCC) is described in Figure 1.10.

The PCRF entity provides to the PCEF entity, integrated in the PGW entity, the necessary information for the control and the charging of the traffic data (IP packets).

This information is stored in the subscription profile repository (SPR) during the creation of the subscription.

Traffic control includes the following:

– association between a service data flow (SDF) and EPS bearer;

– blocking or allowing IP packets;

– allocation of QCI parameter to EPS bearer.

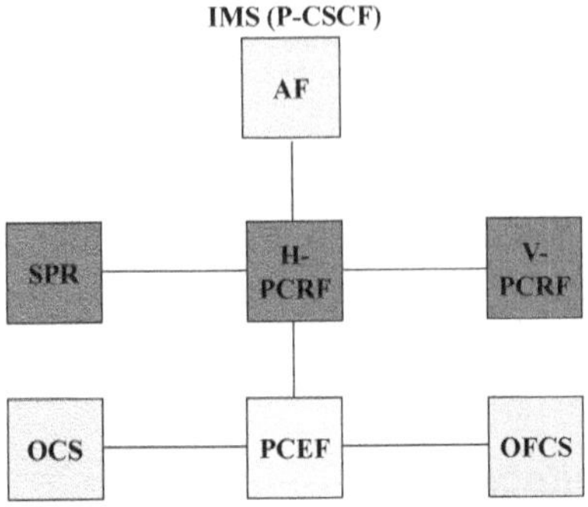

Figure 1.10. *Functional architecture of PCC*

The charging method defines as if the PCEF entity has to obtain credit from the OCS entity for online charging or if it has to generate information submitted to the OFCS entity.

The PCEF entity executes the rules provided by the PCRF entity to control the traffic flow, the accounting of traffic volume and the charging.

The PCEF entity may relate to the PCRF entity a change of state of a service flow, as in the case of loss of radio coverage of the mobile.

The PCRF entity may receive a session request from the AF (Application Function) entity as in the case of the establishment of a voice or conversational video communication initialized at the IMS network.

The PCRF entity may provide the AF entity information about events occurring in the mobile network as in the case of loss of radio coverage of the mobile.

In case of roaming, PCEF entity located in the visited network requests rules to the Visited-PCRF (V-PCRF) entity, which gets them from the Home-PCRF (H-PCRF) entity.

1.5.2. *Protocol architecture*

The Gx interface is the reference point between the PCRF and PCEF entities. This interface supports the DIAMETER signaling.

The Rx interface is the reference point between the PCRF entity and the AF entity, represented by the P-CSCF entity as in the case of the IMS network. This interface supports the DIAMETER signaling.

The Gy interface is the reference point between the PCEF and OCS entities. This interface supports the DIAMETER signaling.

The Gz interface is the reference point between the PCEF and OFCS entities. This interface supports the DIAMETER signaling.

The S9 interface is the reference point located between the H-PCRF entity located in the home network and the V-PCRF entity located in the visited network. This interface supports the DIAMETER signaling.

The Sp interface is the reference point between the PCRF and SPR entities. The protocol used in this interface is not defined.

1.6. DIAMETER routers

DIAMETER agent is a DIAMETER router that can reduce the meshing of DIAMETER sessions between different nodes located, on the one hand, in the entities of EPS or IMS networks, and, on the other hand, in the PCC, OCS and OFCS functions and in the HSS and SLF database (Figure 1.11).

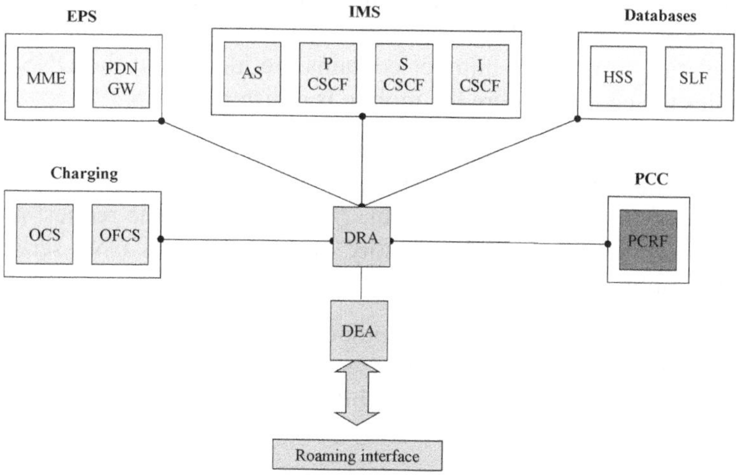

Figure 1.11. *DIAMETER routers*

DIAMETER routing is done on the basis of the operator's domain name to obtain the IP address of the next hop and on the basis of the identity of the destination to obtain the IP address of the DIAMETER node.

DIAMETER routing agent (DRA) only performs routing DIAMETER messages and does not analyze the content of DIAMETER messages. The DRA router is deployed in the home network of the operator.

DIAMETER edge agent (DEA) performs the routing of DIAMETER messages and control of their content according to rules established by the operator. The DEA router is deployed at roaming interfaces between the visited network and the home network.

1.7. ENUM system

The ENUM mechanism allows for using the phone number (E.164 or TEL URI) to determine the network of the called.

The ENUM mechanism is based on domain name system (DNS) resolution which converts the E.164 identity or TEL URI to a SIP URI identity containing the domain name of the destination network.

The ENUM system is structured in three levels of servers:

– level 1: these servers contain pointers to the level-2 servers. The DNS level-1 server manages the *e164enum.net* domain. The response to the request relating to a phone number contains the identity of the DNS server that handles the country where the mobile is registered;

– level 2: these servers have authority over the country code and contain pointers to the level-3 servers. The DNS level-2 servers manage the *<country code> e164enum.net* domain. The response to the request relating to a phone number contains the identity of the operator's DNS server that handles the mobile;

– level 3: these servers have authority over the codes assigned to operators and subscribers. The response to the request relating to a phone number contains the SIP URI of the mobile identity.

1.8. IPX network

The internet protocol exchange (IPX) network provides the interconnection between mobile network operators

A mobile network requires only a single connection to an IPX network to be able to interconnect with other networks of fixed and mobile operators.

IPX network can offer several types of services:

– transport service which is to route IP packets;

– proxy service which is to route DIAMETER messages, to route SIP signaling messages and to switch RTP streams;

– virtual roaming service that allows an operator to replace multiple bilateral roaming agreements by a single agreement with the virtual roaming operator;

– ENUM database service by taking into account the level-1 DNS server that manages the *e164enum.net* domain.

2

Signaling Protocols

2.1. NAS protocol

The non-access stratum (NAS) protocol is the signaling exchanged between the user equipment (UE) and the mobility management entity (MME).

The NAS protocol is transported by the radio resource control (RRC) protocol over the radio interface LTE-Uu and by the S1-AP (Application Part) protocol over the S1-MME interface.

The NAS protocol comprises the following two protocols:

– EPS mobility management (EMM): this takes care of controlling mobility and security;

– EPS session management (ESM): this controls the bearer establishment.

From the point of view of the MME entity, the mobile can be in one of two operational states: EMM-REGISTERED or EMM-DEREGISTERED.

In the EMM-DEREGISTERED state, the mobile location is not known to the MME entity, and therefore, it cannot be contacted.

The switch to the registered state takes place when the mobile attaches, which comprises the following four procedures:

– mutual authentication of the mobile and the MME entity;

– registration of the mobile location with the MME entity;

– assignment of the globally unique temporary identity (GUTI) to the mobile;

– establishment of the default bearer.

The switch to the deregistered state takes place when the mobile detaches or when the attachment of the mobile, the update of its location or the service request are rejected by the MME entity.

2.1.1. *EMM messages*

2.1.1.1. *Attachment and detachment*

The procedure of attachment is initiated by the mobile in the deregistered state, by sending the message EMM ATTACH REQUEST to the MME entity.

This message contains the mobile identity, GUTI or international mobile subscriber identity (IMSI) and the tracking area identity (TAI).

The mobile attaches the message ESM PDN CONNECTIVITY REQUEST to establish the default bearer.

Upon receiving this message, the MME entity begins the authentication and security procedures for the NAS protocol.

If they are successfully completed, the MME entity responds with the message EMM ATTACH ACCEPT, containing a new GUTI, and the message ESM ACTIVATE DEFAULT EPS BEARER CONTEXT REQUEST, to establish the default bearer.

The mobile responds with the message EMM ATTACH COMPLETE containing the message ESM ACTIVATE DEFAULT EPS BEARER CONTEXT ACCEPT, to acknowledge the previous message.

If the procedures are not successful, the MME entity responds with the message EMM ATTACH REJECT, containing the message ESM PDN CONNECTIVITY REJECT, which causes the mobile to detach.

Detachment may be initiated by the mobile or the MME entity by sending the message EMM DETACH REQUEST.

The response EMM DETACH ACCEPT concludes the detachment procedure. The response is not transmitted by the MME entity if the detach request sent by the mobile indicates that it has been turned off. The detachment procedure implicitly causes the release of the active bearers.

2.1.1.2. *Authentication*

The procedure of mutual authentication is initiated by the MME entity by sending the message EMM AUTHENTICATION REQUEST, containing a random number RAND and the authentication network (AUTN).

The mobile uses the RAND received to locally compute its own authentication code RES (Result) and that of the network (AUTN) and compares the AUTN calculated to the one received from the MME entity.

If the MME entity is authenticated, the mobile responds with the message EMM AUTHENTICATION RESPONSE, containing the authentication code RES.

Otherwise, it indicates that network authentication failed with the message EMM AUTHENTICATION FAILURE.

The MME entity compares the RES value received from the mobile with that communicated by the HSS.

If the two codes are the same, the mobile is authenticated and the MME entity triggers security mode for NAS signaling.

Otherwise, the MME entity responds with the message EMM AUTHENTICATION REJECT.

2.1.1.3. *Security mode*

When mutual authentication has been successful, the MME begins putting the NAS signaling in security mode by sending the message EMM SECURITY MODE COMMAND. The integrity of this message is protected.

If the check on the integrity of the message EMM SECURITY MODE COMMAND is positive, the mobile responds with the message EMM SECURITY MODE COMPLETE. All subsequent NAS messages are then encrypted and their integrity is checked.

If the check on the integrity of the message EMM SECURITY MODE COMMAND is negative, the mobile responds with the message EMM SECURITY MODE REJECT.

2.1.1.4. *Tracking area update*

The procedure of tracking area update is periodically initiated so that the mobile can maintain its tracking area or when the mobile has changed its location area. The mobile, in the registered state, sends the message EMM TRACKING AREA UPDATE REQUEST to the MME entity.

The MME entity responds either with the message EMM TRACKING AREA UPDATE ACCEPT if it accepts the update, or else with the message EMM TRACKING AREA UPDATE REJECT, indicating the cause of the rejection.

If the message EMM TRACKING AREA UPDATE ACCEPT attributes a new GUTI to the mobile, the mobile confirms receipt of this by sending the message EMM TRACKING AREA UPDATE COMPLETE.

2.1.1.5. *Service request*

The service request is sent, when the mobile is in idle mode, to re-establish the default bearers on the LTE-Uu and S1-U interfaces.

The service request is initiated by the mobile, by sending the EMM SERVICE REQUEST when signaling or traffic data is waiting.

The mobile is notified of awaiting data at the level of the network by way of the paging procedure.

The MME entity may reject the request, in which case it responds with the message EMM SERVICE REJECT. This response causes the switch of the mobile to the deregistered state.

2.1.2. *ESM messages*

The mobile sends the request to establish the default bearer when the mobile attaches to the MME entity.

The establishment request for the dedicated bearer can be transmitted by the network or the mobile.

The dedicated bearer corresponds to a specific request by the mobile, following for example the establishment of a voice or a conversational video.

The dedicated bearer is associated with a particular quality of service, corresponding to a particular QoS class identifier (QCI) which is different to that of the default bearer.

Table 2.1 summarizes the ESM messages exchanged for the establishment, modification and release of the default bearer and the dedicated bearer.

Establishment of default bearer, initiated by the network		
Source	Message	Destination
MME	ACTIVATE DEFAULT EPS BEARER CONTEXT REQUEST	UE
UE	ACTIVATE DEFAULT EPS BEARER CONTEXT ACCEPT or ACTIVATE DEFAULT EPS BEARER CONTEXT REJECT	MME
Establishment of default bearer, initiated by the mobile		
Source	Message	Destination
UE	PDN CONNECTIVITY REQUEST	MME
MME	ACTIVATE DEFAULT EPS BEARER CONTEXT REQUEST or PDN CONNECTIVITY REJECT	UE
Establishment of dedicated bearer, initiated by the network		
Source	Message	Destination
MME	ACTIVATE DEDICATED EPS BEARER CONTEXT REQUEST	UE
UE	ACTIVATE DEDICATED EPS BEARER CONTEXT ACCEPT or ACTIVATE DEDICATED EPS BEARER CONTEXT REJECT	MME
Modification of dedicated bearer, initiated by the network		
Source	Message	Destination
MME	MODIFY EPS BEARER CONTEXT REQUEST	UE
UE	MODIFY EPS BEARER CONTEXT ACCEPT or MODIFY EPS BEARER CONTEXT REJECT	MME

Release of dedicated bearer, initiated by the network		
Source	Message	Destination
MME	DEACTIVATE EPS BEARER CONTEXT REQUEST	UE
UE	DEACTIVATE EPS BEARER CONTEXT ACCEPT	MME
Establishment of dedicated bearer, initiated by the mobile		
Source	Message	Destination
UE	BEARER RESOURCE ALLOCATION REQUEST	MME
MME	ACTIVATE DEDICATED EPS BEARER CONTEXT REQUEST or ACTIVATE DEDICATED EPS BEARER CONTEXT REJECT or MODIFY EPS BEARER CONTEXT REQUEST	UE
Modification of dedicated bearer, initiated by the mobile		
Source	Message	Destination
UE	BEARER RESOURCE MODIFICATION REQUEST	MME
MME	ACTIVATE DEDICATED EPS BEARER CONTEXT REQUEST or MODIFY EPS BEARER CONTEXT REQUEST or DEACTIVATE EPS BEARER CONTEXT REQUEST or BEARER RESOURCE MODIFICATION REJECT	UE
Release of dedicated bearer, initiated by the mobile		
Source	Message	Destination
UE	PDN DISCONNECT REQUEST	MME
MME	DEACTIVATE EPS BEARER CONTEXT REQUEST or PDN DISCONNECT REJECT	UE

Table 2.1. *ESM messages*

2.2. RRC protocol

The RRC protocol is the signaling exchanged between the mobile and the evolved node base station (eNB) over the LTE-Uu radio interface.

The RRC protocol performs the following functions:

– broadcast of the system information related to the characteristics of the radio interface;

– control of the RRC connection: this procedure includes the paging, the establishment, the modification and the release of the signaling radio bearer (SRB) and the data radio bearer (DRB). It also includes the activation of security mode over the LTE-Uu interface, the procedure for which includes putting mechanisms in place to encrypt the traffic and RRC signaling flows, and to control the integrity of the RRC signaling flows;

– control of handover: this procedure executes the changing of cell between two eNB entities (intra-system handover) or between an eNB entity and a base station from a second- or third-generation mobile network (inter-system handover);

– measurement reporting: the eNB entity can trigger measurements carried out by the mobile, either periodically or on demand, to prepare for handover;

– transport of the NAS messages exchanged between the mobile and the MME entity.

From the point of view of the eNB entity, the mobile may be in one of two operational states: idle mode (RRC_IDLE) or connected mode (RRC_CONNECTED).

In idle mode, the mobile is not known to the eNB entity. It remains in this state until the RRC connection setup procedure is completed. The setup procedure is triggered by the mobile when it wishes to transmit traffic or signaling data, the mobile being used the SRB0 bearer.

In connected mode, the mobile can transmit and receive signaling and traffic data. The mobile is attributed an identifier which is unique to the cell, the cell radio network temporary identity (C-RNTI). In this state, the mobile uses either the SRB1 bearer for RRC messages with possible associated NAS messages, or the SRB2 bearer for RRC messages transporting solely NAS messages.

Table 2.2 summarizes the type of SRB, the mode of radio link control (RLC) protocol and the channels used by the different RRC messages over the radio interface.

<table>
<tr><td colspan="5">MasterInformationBlock</td></tr>
<tr><td>SRB</td><td>RLC mode</td><td>Logical channel</td><td>Transport channel</td><td>Physical channel</td></tr>
<tr><td>Not available</td><td>TM</td><td>BCCH</td><td>BCH</td><td>PBCH</td></tr>
<tr><td colspan="5">SystemInformationBlock</td></tr>
<tr><td>SRB</td><td>RLC mode</td><td>Logical channel</td><td>Transport channel</td><td>Physical channel</td></tr>
<tr><td>Not available</td><td>TM</td><td>BCCH</td><td>DL-SCH</td><td>PDSCH</td></tr>
<tr><td colspan="5">Paging</td></tr>
<tr><td>SRB</td><td>RLC mode</td><td>Logical channel</td><td>Transport channel</td><td>Physical channel</td></tr>
<tr><td>Not available</td><td>TM</td><td>PCCH</td><td>PCH</td><td>PDSCH</td></tr>
<tr><td colspan="5">ConnectionSetup
ConnectionReject
ConnectionReestablishment
ConnectionReestablishmentReject</td></tr>
<tr><td>SRB</td><td>RLC mode</td><td>Logical channel</td><td>Transport channel</td><td>Physical channel</td></tr>
<tr><td>SRB0</td><td>TM</td><td>CCCH</td><td>DL-SCH</td><td>PDSCH</td></tr>
<tr><td colspan="5">ConnectionRequest
ConnectionReestablishmentRequest</td></tr>
<tr><td>SRB</td><td>RLC mode</td><td>Logical channel</td><td>Transport channel</td><td>Physical channel</td></tr>
<tr><td>SRB0</td><td>TM</td><td>CCCH</td><td>UL-SCH</td><td>PUSCH</td></tr>
</table>

ConnectionReconfiguration *ConnectionRelease* *SecurityModeCommand*				
SRB	RLC mode	Logical channel	Transport channel	Physical channel
SRB1	AM	DCCH	DL-SCH	PDSCH
DLInformationTransfer (1)				
SRB	RLC mode	Logical channel	Transport channel	Physical channel
SRB2	AM	DCCH	DL-SCH	PDSCH
ULInformationTransfer (2)				
SRB	RLC mode	Logical channel	Transport channel	Physical channel
SRB2	AM	DCCH	UL-SCH	PUSCH
ConnectionSetupComplete *SecurityModeComplete* *SecurityModeFailure* *ConnectionReconfigurationComplete* *ConnectionReestablishmentComplete* *MeasurementReport*				
SRB	RLC mode	Logical channel	Transport channel	Physical channel
SRB1	AM	DCCH	UL-SCH	PUSCH

Table 2.2. *RRC messages: 1) transport of NAS messages only, downstream; 2) transport of NAS messages only, upstream*

2.2.1. *System information*

The information relating to the radio interface is divided between the messages *MasterInformationBlock* and *SystemInformationBlock*.

These messages are transmitted periodically and a change in these data is notified to the mobile by paging.

The *MasterInformationBlock* message contains the following data:

– the bandwidth of the radio signal for the downstream direction (1.4, 3, 5, 10, 15 and 20 MHz);

– the system frame number (SFN);

– the configuration of the physical channel PHICH of the radio interface. The configuration of this physical channel is defined by the operator.

All *SystemInformationBlock* messages, with the exception of the *SystemInformationBlockType1* message, are mapped in a *SystemInformation* message.

Each *SystemInformation* message contains *SystemInformationBlock* messages with the same periodicity.

The *SystemInformationBlockType2* message must necessarily be mapped in the *SystemInformation1* message.

The *SystemInformationBlockType1* message contains the following data:

– the mobile country code (MCC) and mobile network code (MNC) of the mobile network;

– the mobile network code (TAC) of the location area. The identity of the TAI is a concatenation of the MCC, MNC and TAC codes.

– the E-UTRAN Cell global identifier (ECGI);

– the periodicity of the *SystemInformation* messages and the types of *SystemInformationBlock* messages that they contain.

Table 2.3 shows the data transported by the different types of *SystemInformationBlock* message.

SystemInformationBlockType2	Bandwidth in upstream direction Configuration of physical channels
SystemInformationBlockType3	Cell selection parameters
SystemInformationBlockType4	EPS neighboring cells, same frequency
SystemInformationBlockType5	EPS neighboring cells, different frequency
SystemInformationBlockType6	UMTS neighboring cells
SystemInformationBlockType7	GSM/GPRS neighboring cells
SystemInformationBlockType8	CDMA 2000 neighboring cells
SystemInformationBlockType9	Identity of the home eNB (HeNB)
SystemInformationBlockType10 SystemInformationBlockType11	Earthquake and tsunami warning system (ETWS)
SystemInformationBlockType12	Commercial Mobile Alert System (CMAS)
SystemInformationBlockType13	Information related to the MBSFN (MBMS over Single Frequency Network) area

Table 2.3. *SystemInformationBlock messages*

2.2.2. *Control of RRC connection*

The different procedures associated with the control of the RRC connection relate to paging, RRC connection setup, activation of security mode, RRC connection reconfiguration, RRC connection re-establishment and RRC connection release.

The *Paging* message is used by the eNB entity to alert one or more mobiles in the RRC_IDLE state.

The *Paging* message also helps to inform the mobile in RRC_IDLE or RRC_CONNECTED state about a change in the system information or about a notification on ETWS transmitted in the *SystemInformationBlockType10* and *SystemInformationBlockType11* messages or CMAS transmitted in the *SystemInformationBlockType12* message.

The RRC *ConnectionRequest* message is used by the mobile to request the establishment of an RRC connection.

The RRC *ConnectionSetup* message is used by the eNB entity to configure the SRB1 bearer.

The RRC *ConnectionSetupComplete* message is used by the mobile to confirm the setup of the RRC connection. This message can also transport NAS messages.

The RRC *ConnectionReject* message is used by the eNB entity to reject the request for the RRC connection.

Upon receiving the context about the mobile from the MME entity, the eNB entity activates security mode for the RRC messages.

The *SecurityModeCommand* message is used by the eNB entity to command the activation of security mode on the radio interface.

The *SecurityModeCommand* message is only checked for integrity.

The *SecurityModeComplete* message is used by the mobile to confirm the activation of security mode.

The *SecurityModeFailure* message is used by the mobile to indicate that security mode was unable to be activated.

The encryption of the RRC messages will be effective only if the procedure has been successful.

Having initiated the security mode activation procedure, the eNB entity begins the activation of the DRB. The RRC messages are encrypted and checked for integrity.

The RRC *ConnectionReconfiguration* message is used by the eNB entity to command a modification of the RRC connection. This message may relate to the configuration of the measurements, control of the mobility and configuration of the DRB default bearer. This message can also transport NAS messages.

The RRC *ConnectionReconfigurationComplete* message is used by the mobile to confirm the reconfiguration of the RRC connection.

The RRC *ConnectionReestablishmentRequest* message is used by the mobile to request the re-establishment of the RRC connection.

The RRC *ConnectionReestablishment* message is used by the eNB entity to re-establish the RRC connection.

The RRC *ConnectionReestablishmentComplete* message is used by the mobile to confirm the re-establishment of the RRC connection.

The RRC *ConnectionReestablishmentReject* message is used by the eNB entity to indicate that the reestablishment of the RRC connection has been rejected.

The RRC *ConnectionRelease* message is used by the eNB entity to release the RRC connection. The procedure can also be used to redirect the mobile to a different frequency band. In exceptional cases, the mobile can terminate the RRC connection without alerting the eNB entity.

2.2.3. *Measurement report*

The measurements carried out by the mobile must be in line with the configuration indicated in the RRC *ConnectionReconfiguration* message transmitted by the eNB entity.

The mobile sends the eNB entity the measurements in the RRC *MeasurementReport* message.

Measurements carried out on the serving cell and neighboring cells are used for the selection of the cell and for the handover.

Intra-frequency measurement, essential for mobility, is configured when connecting.

Inter-frequency and inter-radio access technology (RAT) measurement can also be configured when connecting.

As the mobile does not generally have several radio receivers, the inter-frequency and inter-RAT measurement should be performed at intervals arranged in the frame.

The configuration of measurements to be performed by the mobile is triggered by the eNB entity in RRC *ConnectionSetup*, *Connection Reconfiguration*, *ConnectionReestablishment* messages.

The measurement configurations to perform define the following parameters:

– the object that identifies the radio channel;

– the event triggering the measurement report;

– the combination of objects and events;

– the measurement of filtering parameters;

– the periodicity of measurements.

2.3. S1-AP protocol

The S1-AP protocol is the signaling exchanged between the eNB and MME entities over the S1-MME interface.

The S1-AP protocol performs the following functions:

– activation of the context of the mobile by the MME entity;

– establishment, modification and release of the EPS radio access bearer (E-RAB);

– handover management;

– paging. This procedure tells the eNB entity that the message needs to be broadcast in the cell;

– transport of the NAS signaling exchanged between the mobile and the MME entity;

– establishment of the S1-MME interface.

Table 2.4 summarizes the S1-AP messages sent to paging, context management, bearer management, mobility management, management of the S1-MME interface and NAS messages transport.

Function	*Request*	*Response*
Context management	INITIAL CONTEXT SETUP REQUEST	INITIAL CONTEXT SETUP RESPONSE or INITIAL CONTEXT SETUP FAILURE
	UE CONTEXT RELEASE COMMAND	UE CONTEXT RELEASE COMPLETE
Function	*Request*	*Response*
Bearer management	E-RAB SETUP / MODIFY REQUEST	E-RAB SETUP / MODIFY RESPONSE
	E-RAB RELEASE COMMAND	E-RAB RELEASE RESPONSE
	E-RAB RELEASE INDICATION	None
Function	*Request*	*Response*
Paging	PAGING	None
Function	*Request*	*Response*
Mobility management	HANDOVER REQUIRED	HANDOVER COMMAND
	HANDOVER REQUEST	HANDOVER REQUEST ACKNOWLEDGE or HANDOVER FAILURE
	eNB STATUS TRANSFER	None
	MME STATUS TRANSFER	None
	HANDOVER NOTIFY	None
	PATH SWITCH REQUEST	PATH SWITCH ACKNOWLEDGE or PATH SWITCH FAILURE

Function	*Request*	*Response*
Management of the S1-MME interface	S1 SETUP REQUEST	S1 SETUP RESPONSE or S1 SETUP FAILURE
	ENB CONFIGURATION UPDATE	ENB CONFIGURATION UPDATE ACKNOWLEDGE or ENB CONFIGURATION UPDATE FAILURE
	MME CONFIGURATION UPDATE	MME CONFIGURATION UPDATE ACKNOWLEDGE or ENB CONFIGURATION UPDATE FAILURE
	OVERLOAD START	None
	OVERLOAD STOP	None
Function	*Request*	*Response*
Transport of NAS signaling	INITIAL UE MESSAGE	None
	DOWNLINK NAS TRANSPORT	None
	UPLINK NAS TRANSPORT	None

Table 2.4. *S1-AP messages*

2.3.1. *Context management*

The context of the mobile has to be established at the level of the eNB and MME entities so as to transmit the mobile traffic and the NAS signaling as well.

The context of the mobile includes the contexts relating to the default bearer, security, capacities of the mobile and roaming restrictions.

Context setup for the mobile begins with the INITIAL CONTEXT SETUP REQUEST message transmitted by the MME entity to the eNB entity. This message follows the reception of the INITIAL UE MESSAGE message.

The MME has to prepare the establishment of the default bearer before receiving the INITIAL CONTEXT SETUP RESPONSE message. This message might contain the cause of the failure to set up the context of the mobile, such as the lack of radio resources.

If the eNB is unable to establish the context of the mobile, it responds with the INITIAL CONTEXT SETUP FAILURE message.

Release of the context of the mobile is done by way of the UE CONTEXT RELEASE COMMAND message transmitted by the MME entity to the eNB entity, for instance when the mobile changes cell.

This message is acknowledged in return by the UE CONTEXT RELEASE COMPLETE response.

2.3.2. *Bearer management*

Establishment and modification of the E-RAB dedicated bearer are initiated by the MME entity by sending the E-RAB SETUP/MODIFY REQUEST messages.

The eNB entity responds positively or negatively by sending the E-RAB SETUP/MODIFY RESPONSE messages.

Release of the dedicated bearer is initiated by the MME entity by sending the E-RAB RELEASE COMMAND message or by the eNB entity by sending an E-RAB RELEASE INDICATION message. Upon receiving this message, the MME entity begins the procedure of release of the dedicated bearer.

2.3.3. *Mobility management*

The decision regarding handover based on the S1 interface is made by the source eNB entity.

The phase of handover preparation begins with the sending of the HANDOVER REQUIRED message to the MME entity.

When the reservation of resources by the target eNB entity is effective, the MME entity responds with the HANDOVER COMMAND message.

The MME entity directs the target eNB entity to reserve the radio resources using the HANDOVER REQUEST message.

If the operation is successful, the target eNB entity responds with the HANDOVER REQUEST ACKNOWLEDGE message. This message can contain the elements for construction of a GTP-U tunnel to transfer the received data from the source eNB entity to the target eNB entity so they can be transmitted to the mobile.

If not, the target eNB entity responds with the HANDOVER FAILURE message.

The source eNB entity has to transfer the value of the field sequence number (SN) of the packet data convergence protocol (PDCP) to the target eNB entity to preserve the continuity of the PDCP frame numbering. This operation is done by the transmission of the following messages:

– eNB STATUS TRANSFER from the source eNB entity to the MME entity;

– MME STATUS TRANSFER from the MME entity to the target eNB entity.

This procedure applies only to bearers who use the acknowledgment mode (AM) of the RLC protocol, which is not the case of a dedicated bearer for voice or video.

When the execution of the handover has been completed, the target eNB entity advises the MME entity of this by way of the HANDOVER NOTIFY message.

Regarding handover based on the X2 interface the PATH SWITCH REQUEST message is transmitted by the target eNB entity to the MME entity for the transfer of the extremity of the GTP-U tunnel corresponding to the source eNB entity to the target eNB entity.

The MME responds with the PATH SWITCH ACKNOWLEDGE message if the response is positive or with PATH SWITCH FAILURE message if not.

2.3.4. *S1-MME interface management*

The eNB entity takes the initiative to activate the S1-MME interface by transmitting the S1 SETUP REQUEST message, indicating the list of serving location area.

The S1 SETUP RESPONSE message from MME entity contains information relating to the MME entity such as its code number, the pool number to which it belongs and the MNC and MCC codes.

The MME entity may respond negatively with the S1 SETUP FAILURE message.

Updates to the information about the eNB entity (or the MME entity respectively) are transmitted by the ENB CONFIGURATION UPDATE message (or the MME CONFIGURATION UPDATE message respectively).

These messages receive a positive response with the ENB/MME CONFIGURATION UPDATE ACKNOWLEDGE messages or a negative one with the ENB/MME CONFIGURATION UPDATE FAILURE messages.

The MME entity notifies the eNB entity of the beginning (or the end respectively) of a state of overload by the OVERLOAD START message (or OVERLOAD STOP message respectively) so as to avoid being selected for the attachment of a new mobile.

2.4. X2-AP protocol

The X2-AP protocol is the signaling exchanged between two eNB entities over the X2 interface.

The X2-AP protocol performs the following functions:

– mobility management: this function enables the source eNB entity to transfer the connection of a mobile to the target eNB entity;

– load management: this function is used by the eNB entities to provide an indication of the load of the cells that they serve;

– X2 interface management: this function is used for the activation of the X2 interface, the reconfiguration and re-initialization of the X2 interface.

Table 2.5 summarizes the X2-AP messages exchanged for mobility management, load management and X2 interface management.

Function	*Request*	*Response*
Mobility management	HANDOVER REQUEST	HANDOVER REQUEST ACKNOWLEDGE or HANDOVER PREPARATION FAILURE
	SN STATUS TRANSFER	None
	UE CONTEXT RELEASE	None
	HANDOVER CANCEL	None
Function	*Request*	*Response*
Load management	LOAD INFORMATION	None
	RESOURCE STATUS REQUEST	RESOURCE STATUS RESPONSE or RESOURCE STATUS FAILURE
	RESOURCE STATUS UPDATE	None
Function	*Request*	*Response*
X2 interface management	X2 SETUP REQUEST	X2 SETUP RESPONSE or X2 SETUP FAILURE
	ENB CONFIGURATION UPDATE	ENB CONFIGURATION UPDATE ACKNOWLEDGE or ENB CONFIGURATION UPDATE FAILURE
	RESET REQUEST	RESET RESPONSE

Table 2.5. *X2-AP messages*

2.4.1. *Mobility management*

The function of mobility management contains the following elementary procedures:

– handover preparation;

– transfer of the state of the SN field of the PDCP protocol;

– deactivation of the context of the mobile;

– handover cancellation.

The procedure of handover preparation is initiated by the source eNB entity by transmission of the HANDOVER REQUEST message to the target eNB entity.

The target eNB reserves the resources and responds with the HANDOVER REQUEST ACKNOWLEDGE message.

This message contains the value of the tunnel endpoint identifier (TEID) used by the GTP-U protocol for the traffic transferred by the source eNB entity to the target eNB entity.

If the resources are unavailable, the target eNB entity sends back the HANDOVER PREPARATION FAILURE message.

The procedure of transfer of the state of the SN field consists of transferring to the eNB entity the value of the SN of the PDCP protocol with the SN STATUS TRANSFER message. At the source eNB entity, this message stops the attribution of the SN of the PDCP protocol for the downstream direction.

The procedure for context release of the mobile is initiated by the target eNB entity by sending the UE CONTEXT RELEASE message to the source eNB entity. Upon receiving this message, the eNB entity eliminates the context of the mobile.

The procedure for cancelling the handover is initiated by the source eNB entity with the HANDOVER CANCEL message. This message causes the target eNB entity to release the resources on the radio interface.

2.4.2. *Load management*

The function of load management includes the following elementary procedures:

– cell-load indication;

– initialization of resource status reports;

– resource status reporting.

The procedure for indication of the load of the cell is initiated by either of the eNB entities with the LOAD INFORMATION message. This message may contain the following elements of information:

– interference overload indication (IOI): this information relates to the interference detected by the eNB entity, for the upstream direction. The eNB entity receiving this information has to decrease the interference transmitted by the mobile;

– high interference indication (HII): this information relates to the interference detected by the eNB entity, for the upstream direction, indicating which bandwidths are affected. The eNB entity receiving this information needs to avoid using the said bandwidth, for the upstream direction, for the mobiles located on the periphery of the cell;

– relative narrowband Tx power (RNTP): this information relates to a decrease in the power transmitted by an eNB entity. The eNB entity receiving this information includes it in its traffic management mechanism.

The procedure for initialization of resource status reporting is initiated by either of the eNB entities with the RESOURCE STATUS REQUEST message.

The eNB receiving this message responds with the RESOURCE STATUS RESPONSE message, which may contain status information for the radio resources, the S1 interface and the load of the eNB entity.

The eNB entity may respond with the RESOURCE STATUS FAILURE message if the reports cannot be generated.

The resource status report is then transmitted periodically by the eNB entity by sending the RESOURCE STATUS UPDATE message.

2.4.3. *X2 interface management*

The X2 interface is set up with the intention of exchanging the configuration data necessary for both eNB entities to function correctly.

One of the eNB entities initiates the procedure by indication of the cells served in the X2 SETUP REQUEST message to a candidate eNB entity.

The candidate eNB entity responds with the X2 SETUP RESPONSE message, also containing the list of serving cells.

The information communicated may also include the list of neighboring cells and the number of antennas for each serving cell.

The eNB entity may refuse the establishment of the X2 interface by sending the X2 SETUP FAILURE message in response.

The X2 setup is followed by configuration updating of the eNB entity if the configuration of the eNB entity changes. The configuration update is initiated by the ENB CONFIGURATION UPDATE message.

The remote eNB entity may respond positively with the CONFIGURATION UPDATE ACKNOWLEDGE message or negatively with the ENB CONFIGURATION UPDATE FAILURE message.

The reset of the X2 interface is intended to align the resources of the eNB entities in the case of an unexpected breakdown. The procedure is initiated by the RESET REQUEST message.

The receiving eNB entity responds with the RESET RESPONSE message. The procedure does not affect the data exchanged during the X2 setup or the configuration update of the eNB entity.

2.5. GTPv2-C protocol

GTP-U (GPRS Tunnel Protocol User) tunnels are used between two entities of the EPS network. Such tunnels enable the traffic data to be compartmentalized. GTP-U traffic tunnels are constructed on the S1-U, S5 and X2 interfaces.

The tunnel is identified by the TEID parameter, the IP addresses and the UDP port numbers. The entity receiving the traffic or signaling data determines the value of the TEID parameter which the sending entity has to use.

The values of the TEID parameter of the GTP-U protocol are exchanged via the GTPv2-C (GPRS Tunnel Protocol Control), S1-AP and X2-AP protocols.

The TEID parameter used for the signaling exchanged over the S5 interface is unique. The same parameter is used for all signaling messages relating to the activation of the various S5 bearers for the different mobiles.

The TEID parameter used for the signaling exchanged over the S10 and S11 interfaces is unique for each mobile. The same parameter is used for all signaling messages relating to the establishment of the various S1-U bearers for the same mobile.

Table 2.6 summarizes the GTPv2-C messages exchanged for the management of support and mobility.

Type of messages	*Request*	*Response*
Bearer management	CREATE / DELETE SESSION REQUEST	CREATE / DELETE SESSION RESPONSE
	CREATE / MODIFY / DELETE BEARER REQUEST	CREATE / MODIFY / DELETE BEARER RESPONSE
	DOWNLINK DATA NOTIFICATION	DOWNLINK DATA NOTIFICATION ACKNOWLEDGE or DOWNLINK DATA NOTIFICATION FAILURE INDICATION
	CREATE / DELETE INDIRECT DATA FORWARDING TUNNEL REQUEST	CREATE / DELETE INDIRECT DATA FORWARDING TUNNEL RESPONSE
Type of messages	*Request*	*Response*
Mobility management	FORWARD RELOCATION REQUEST	FORWARD RELOCATION RESPONSE
	FORWARD RELOCATION NOTIFICATION	FORWARD RELOCATION ACKNOWLEDGE
	FORWARD ACCESS CONTEXT NOTIFICATION	FORWARD ACCESS CONTEXT ACKNOWLEDGE
	CONTEXT REQUEST	CONTEXT RESPONSE
	CONTEXT ACKNOWLEDGE	

Table 2.6. *GTPv2-C messages*

2.5.1. *Bearer management*

The signaling bearer related to a mobile is created by the CREATE SESSION REQUEST message. It is reinforced by the use of a TEID parameter. The message is transmitted:

– by the MME entity to the serving gateway (SGW), over the S11 interface;

– by the target SGW entity for the PDN gateway (PGW), over the S5 interface. The request is transmitted when any of the following procedures are initiated:

- attachment of the mobile,
- traffic request from the mobile,
- updating of the tracking area code (TAC),
- handover.

The SGW entity (or respectively PGW entity) responds to the MME entity (or respectively SGW entity) with the CREATE SESSION RESPONSE message.

The signaling bearer is deactivated by the exchange of the DELETE SESSION REQUEST/RESPONSE messages.

The procedure is triggered when the mobile detaches, when the traffic is released, when the TAC changes, leading to a modification of the SGW entity, or when handover occurs, with a switch of the SGW.

The dedicated bearer specific to a mobile is created similarly, modified possibly and deleted by the exchange of the following messages:

– CREATE / MODIFY / DELETE BEARER REQUEST;

– CREATE / MODIFY / DELETE BEARER RESPONSE.

The DOWNLINK DATA NOTIFICATION message is sent by the SGW entity to the MME entity, over the S11 interface.

The procedure follows the reception by the SGW entity of data from the PGW entity, with the mobile in ECM-IDLE mode. Just after receiving this message, the MME entity sends the S1-AP PAGING message to the eNB entities belonging to the TAC.

The MME entity may respond with the DOWNLINK DATA NOTIFICATION ACKNOWLEDGE message, indicating whether or not the request is accepted or with the DOWNLINK DATA NOTIFICATION FAILURE INDICATION message if the mobile does not respond to the paging or if the mobile service request is rejected.

The CREATE INDIRECT DATA FORWARDING TUNNEL REQUEST/RESPONSE messages create a specific traffic bearer when handover occurs. This bearer forwards the data traffic received by the source eNB entity to the SGW entity to then be re-transmitted to the mobile via the target eNB entity.

2.5.2. *Mobility management*

Mobility management messages are exchanged between the source and target MME entities, when the handover of the mobile imposes a switch of MME entity.

The source MME entity sends the target MME entity the FORWARD RELOCATION REQUEST message containing the context of the mobile.

The target MME entity responds with the FORWARD RELOCATION RESPONSE message when the resources needed for the handover have been reserved.

The response contains the values of the TEID parameter, which will enable the source SGW entity to redirect the traffic to the target SGW entity during handover.

Upon receiving this message, the source MME entity sends the source eNB entity the command to initiate handover.

The source MME entity sends the target MME entity the FORWARD ACCESS CONTEXT NOTIFICATION message to provide it with the elements of the context of the E-RAB bearer, such as the PDCP sequence number.

The target MME entity sends the source MME entity the FORWARD RELOCATION NOTIFICATION message to indicate that the handover procedure is complete.

The new MME entity sends the CONTEXT REQUEST message to the former one in the procedure of TAI updating, to retrieve information about the context of the mobile.

The former MME entity provides this information in the CONTEXT RESPONSE message, which may contain a positive or negative response. The new MME entity acknowledges this previous message with the message CONTEXT ACKNOWLEDGE.

2.6. SIP protocol

Session initiation protocol (SIP) is a control protocol which can establish, modify and terminate multimedia sessions. Media can be added to or removed from an existing session.

SIP is based on a request/response pair such as hypertext transfer protocol (HTTP). Each transaction consists of a request, which uses a particular method and one or more responses.

2.6.1. *Requests*

The request begins with a line containing the method, a uniform resource iidentifier (URI) and the version of the protocol.

```
INVITE sip:bob@homeB.net SIP/2.0
```

2.6.1.1. *REGISTER method*

The REGISTER method is used by an user agent (UA) to notify the REGISTRAR entity of the correspondence between the IP address of the UA entity and URI concerned. This correspondence is necessary for incoming calls.

The use of the URI of the REGISTER request, and of the headers `To` and `From`, is slightly different to that of other requests.

The URI contains only the domain of the REGISTRAR entity, with no part relating to the user.

The header To contains the URI of the UA entity which needs to be registered. The header From contains the URI of the entity performing the registration. This entity is generally identical to that of the header To.

A UA entity may receive a redirect (3xx) or failure (4xx) response, whose header Contact contains the place where the registrations need to be sent.

2.6.1.2. *INVITE method*

The INVITE method is used by a user agent client (UAC) to establish a dialog or a session. The definitive (positive or negative) responses need to be acknowledged by the ACK request.

The INVITE request may contain a message body describing the media that the UAC entity wants to establish. If this description is absent, it needs to be added to the ACK request.

The response 200 OK contains the description of the media that the user agent server (UAS) wants to establish. The media are established with the response 200 OK (if the INVITE request contains the description of the media) or with the ACK request (otherwise).

A successful INVITE request establishes a dialog between two UA entities, which continues until a BYE request is sent by one of the parties to terminate the session.

The dialog is identified by the header Call-ID and the parameter tag of the headers To (parameter specified by the caller) and From (parameter specified by the callee). The headers To and From are respectively specified with the identities of the callee and the caller.

Within the dialog, it is possible for each UA entity to transmit one or more requests, with each request initializing a transaction.

It is possible to transmit a new INVITE request (reINVITE) within a dialog, for an established session, to update the characteristics of the media. If the reINVITE request is declined, the session continues with the previous characteristics.

2.6.1.3. *ACK method*

The ACK method is used to acknowledge a definitive response (2xx, 3xx, 4xx, 5xx and 6xx) to the INVITE request. The ACK request may contain a message body describing the media if this description is not given in the INVITE request.

2.6.1.4. *BYE method*

The BYE method is used to terminate an established session. A session is considered to be established when the response 2xx is received following the INVITE request.

The BYE request is transmitted by either of the UA entities participating in the session. The UAS sends the response back 2xx if the dialog is known, or else sends the response 481 Dialog/Transaction Does Not Exist.

2.6.1.5. *CANCEL method*

The CANCEL method is used to terminate a session which has not yet been successful. It is generated when a provisional response 1xx has been received, but not a definitive response.

Upon receiving a CANCEL request, the PROXY SERVER transmits the request to the next hop (a PROXY SERVER or a UA entity) and confirms the cancellation directly to the source with the response 200 OK.

A UAS receiving the CANCEL request confirms cancellation with the response 200 OK and terminates the dialog initiated by the INVITE request with a definitive negative response 487 Request Terminated.

2.6.1.6. *PRACK method*

The PRACK method is used to acknowledge the reception of a provisional response (1xx), with the exception of the response 100 Trying.

The PRACK request is transmitted when the provisional response received contains the headers CSeq and Rseq. The PRACK request must include the header RAck containing the values of the headers CSeq and Rseq of the provisional response received.

The provisional response is transmitted at the expiration of a timer until the reception of the PRACK request.

The PRACK request is transmitted at the expiration of a timer until the reception of a `200 OK` response.

2.6.1.7. *UPDATE method*

The UPDATE method modifies the characteristics of the media in a session which has not yet been successfully established.

If the session has successfully been established (the INVITE request has received a `2xx` response), the modification of the characteristics of the media is negotiated by the INVITE method (reINVITE).

2.6.1.8. *SUBSCRIBE method*

The SUBSCRIBE method is used when an UA entity wishes to subscribe to a service whereby he would receive event notifications. The type of event is described in the header `Event`.

The entity which accepts the subscription returns the response `200 OK` containing the duration of the subscription in the header `Expires`. The subscription has to be renewed by transmitting a new SUBSCRIBE request. In the absence of renewal, the subscription terminates automatically.

2.6.1.9. *NOTIFY method*

The NOTIFY method enables an entity to notify the occurrence of an event.

This entity needs to receive a response `200 OK` which gives the assurance that the request has been properly received.

Receipt of the response `481 Dialog/Transaction Does Not Exist` automatically terminates the subscription.

2.6.1.10. *REFER method*

The REFER method can be used to transfer media established between two UA entities (e.g. Alice and Bob) to someone else. The REFER request is sent by Alice (transferor) to Carol to resume communication. The header `Refer-to` of the request contains the SIP URI of Bob (transferee). It should be noted that in this scenario, communication is not established between Alice and Carol.

In a second scenario, Alice establishes an additional communication with Carol. Alice then sends the REFER request to Bob to transfer the communication that she has established with Carol. When Bob notifies her that the transfer has been successful, Alice releases the communication established with Bob.

2.6.1.11. *MESSAGE method*

The MESSAGE method is used to transmit short message service (SMS), contained in a message body, between the two UA entities involved. This message is acknowledged by a response `200 OK`. The response must not contain the message body. If the recipient wishes to respond, he must in turn generate a new MESSAGE request.

2.6.2. *Responses*

The response begins with a line containing the version of the protocol, followed by a code for the type of response and a textual description of the code.

```
SIP/2.0 200 OK
```

The different types of responses are detailed in Tables 2.7 to 2.13.

Type of response	*Description*
`1xx`	Provisional response
`2xx`	Definitive positive response
`3xx`	Definitive redirect response
`4xx`	Definitive negative response, error due to client
`5xx`	Definitive negative response, error due to network
`6xx`	Definitive negative response, global error

Table 2.7. *Types of respones*

Response	*Description*
100 Trying	The sender is informed that the SIP message has been received.
180 Ringing	The caller is informed that the callee is alerted to an incoming call by a ringing tone.
181 Call Is Being Forwarded	The caller is informed that its call has been transferred to a different recipient.
183 Session Progress	The caller is informed that its call is being processed.

Table 2.8. *1xx-type responses*

Response	*Description*
200 OK	The response acknowledges receipt of the request.
202 Accepted	The callee has received the request, requiring a different treatment thereafter

Table 2.9. *2xx-type responses*

Response	*Description*
300 Multiple Choices	The redirect indicates multiple contacts, the order of which is significant.
301 Moved Permanently	The redirect is permanent.
302 Moved Temporarily	The redirect is temporary.
305 Use Proxy	The redirect is performed to a PROXY SERVER.
380 Alternative Service	The redirect points to a different entity (e.g. voicemail service).

Table 2.10. *3xx-type responses*

Response	*Description*
`401 Unauthorized`	The REGISTER request requires authentication
`486 Busy Here`	The callee is busy
`487 Request Terminated`	The response is transmitted by the UA when it receives a CANCEL request

Table 2.11. *4xx-type responses*

Response	*Description*
`500 Server Internal Error`	An internal error has occurred on the PROXY SERVER or the REGISTRAR, so the request has to be repeated later.
`502 Bad Gateway`	The gateway to a different network has detected a fault.
`503 Service Unavailable`	The service is temporarily unavailable.
`504 Gateway Timeout`	The gateway to a different network cannot relay the request because a timer has run out.
`505 Version Not Supported`	The request is denied because of the version of the SIP.
`513 Message Too Large`	The request is denied because the size of the SIP message is too great.

Table 2.12. *5xx-type responses*

Response	*Description*
`600 Busy Everywhere`	The network is saturated.
`603 Decline`	The call is refused.
`604 Does Not Exist Anywhere`	The SIP URI of the callee does not exist anywhere.
`606 Not Acceptable`	Some aspects of the session are not acceptable.

Table 2.13. *6xx-type responses*

2.7. SDP protocol

The session description protocol (SDP) gives a description of the flow for which the establishment of the session is implemented by the SIP. The SDP message constitutes the message body attached to the SIP message. It, generally appears in the INVITE request and in the response `200 OK`.

The parameters which characterize the media flows are as follows:

– the type of media (audio, video, data);

– the transport protocol (e.g. RTP);

– the format of the media (e.g. the type of codec for voice and video);

– the IP address to which the media need to be transmitted;

– the number of the destination port.

The SDP message is a set of lines of code in the format `<type>=<value>`. The field `<type>` contains a character (Table 2.14). The content of the field `<value>` depends on the type.

Field <type>	*Description*
v	Version of the protocol `<value> = <0>`
o	Identifier of the origin and of the session
s	Name of the session `<value> = <->`
c	Information about the connection
t	Activity time of the session `<valeur> = <0 0>`
m	Description of the media
a	Complementary attribute of the media

Table 2.14. *Structure of SDP message*

2.8. DIAMETER protocol

DIAMETER protocol is used by evolved packet system (EPS), IMS and policy and charging control (PCC) networks to ensure authentication, authorization and accounting.

2.8.1. *Application to EPS network*

Table 2.15 summarizes the DIAMETER messages exchanged over the S6a interface between the MME and HSS entities.

Request	*Response*
Authentication-Information-Request (AIR)	Authentication-Information-Answer (AIA)
Update-Location-Request (ULR)	Update-Location-Answer (ULA)
Purge-UE-Request (PUR)	Purge-UE-Answer (PUA)
Insert-Subscriber-Data-Request (IDR)	Insert-Subscriber-Data-Answer (IDA)
Cancel-Location-Request (CLR)	Cancel-Location-Answer (CLA)
Delete-Subscriber-Data-Request (DDR)	Delete-Subscriber-Data-Answer (DDA)
Reset-Request (RSR)	Reset-Answer (RSA)
Notify-Request (NOR)	Notify-Answer (NOA)

Table 2.15. *DIAMETER messages over S6a interface*

AIR and AIA messages allow an MME entity to retrieve the authentication data of the mobile (RAND, RES, AUTN and K_{ASME}).

ULR and ULA messages allow:

– to provide the HSS entity with the identity of the MME entity who attached the mobile and information on the mobile;

– the MME entity to retrieve the profile mobile data.

PUR and PUA messages are used to inform the HSS entity that MME entity deleted the mobile profile, after a long period of inactivity.

CLR and CLA messages allow the HSS entity to delete the mobile profile stored in the MME entity.

IDR and IDA messages allow the HSS entity updating mobile profile stored in the MME entity.

DDR and DDA messages allow the HSS entity to remove elements of the mobile profile stored in the MME entity.

RSR and RSA messages allow informing the MME entity of a restart of the HSS entity.

NOR and NOA messages allow to inform the HSS entity on certain events on the mobile.

2.8.2. *Application to IMS network*

Table 2.16 summarizes DIAMETER messages exchanged over Cx interface between, on the one hand, serving call session control function (S-CSCF) or interrogating-CSCF (I-CSCF), while on the other hand, the HSS entity.

The same DIAMETER messages are exchanged over the Dx interface between, on the one hand, the S-CSCF or I-CSCF entities, and, on the other hand, the subscription locator function (SLF).

Request	*Response*
Multimedia-Authentication-Request (MAR)	Multimedia-Authentication-Answer (MAA)
Server-Assignment-Request (SAR)	Server-Assignment-Answer (SAA)
Registration-Termination-Request (RTR)	Registration-Termination-Answer (RTA)
Push-Profile-Request (PPR)	Push-Profile-Answer (PPA)
User-Authorization-Request (UAR)	User-Authorization-Answer (UAA)
Location-Information-Request (LIR)	Location-Information-Answer (LIA)

Table 2.16. *DIAMETER messages over Cx interface*

MAR and MAA messages are exchanged between S-CSCF and HSS entities and allow the S-CSCF entity to retrieve the authentication data of the mobile.

SAR and SAA messages are exchanged between S-CSCF and HSS entities and allow the S-CSCF entity to retrieve the mobile profile.

RTR and RTA messages are exchanged between S-CSCF and HSS entities which, in turn, allow the HSS entity to deregister the mobile.

PPR and PPA messages are exchanged between S-CSCF and HSS entities and allow the HSS entity to update the profile data of the mobile.

UAR and UAA messages are exchanged between I-CSCF and HSS entities and allow the I-CSCF entity to retrieve the list of S-CSCF entities that can register the mobile.

LIR and LIA messages are exchanged between HSS and I-CSCF entities and allow the I-CSCF entity to retrieve the address of the S-CSCF entity that has registered the mobile.

Table 2.17 summarizes DIAMETER messages exchanged over Sh interface between telephony application server (TAS) and the HSS entity.

The same DIAMETER messages are exchanged over the Dh interface between the TAS and SLF entities.

Request	*Response*
User-Data-Request (UDR)	User-Data-Answer (UDA)
Profile-Update-Request (PUR)	Profile-Update-Answer (PUA)
Subscribe-Notifications-Request (SNR).	Subscribe-Notifications-Answer (SNA)
Push-Notification-Request (PNR)	Push-Notification-Answer (PNA)

Table 2.17. *DIAMETER messages over Sh interface*

UDR and UDA messages allow the TAS entity to retrieve profile data stored in the HSS entity.

PUR and PUA messages allow the TAS entity to update the profile data stored in the HSS entity.

SNR and SNA messages allow the TAS entity to subscribe to notifications relating to modifications of profile data stored in the HSS entity.

PNR and PNA messages allow the TAS entity to receive the notification of modifications in the profile data stored in the HSS entity.

2.8.3. *Application to PCC function*

Table 2.18 summarizes DIAMETER messages exchanged over the Gx interface between policy charging and rules function (PCRF) and policy and charging enforcement function (PCEF) hosted in the PGW entity.

Request	*Response*
Credit-Control-Request (CCR)	Credit-Control-Answer (CCA)
Re-Auth-Request (RAR)	Re-Auth-Answer (RAA)

Table 2.18. *DIAMETER messages over Gx interface*

CCR and CCA messages enable the PCEF entity to solicit the PCRF to:

– retrieve the rules to apply to the default bearer created by the EPS network;

– inform the PCRF entity to the termination of the session on the EPS network.

RAR and RAA messages allow the PCRF entity to solicit the PCEF entity in order to provide the rules to be applied for the dedicated bearer.

Table 2.19 summarizes DIAMETER messages exchanged over the Rx interface between the PCRF and AF (Application Function) entities.

Request	*Response*
Authenticate and Authorize Request (AAR)	Authenticate and Authorize Answer (AAA)
Re-Auth-Request (RAR)	Re-Auth-Answer (RAA)
Session Termination Request (STR)	Session Termination Answer (STA)
Abort-Session-Request (ASR)	Abort-Session-Answer (ASA)

Table 2.19. *DIAMETER messages over Rx interface*

AAR and AAA messages allow the AF entity to provide the characteristics of the media so that the EPS network can establish the dedicated bearer.

RAR and RAA messages allow the PCRF entity to notify the AF entity that a number of IP flows were disabled following a deactivation applied at the PCEF entity.

STR and STA messages allow the AF entity to inform the session is finished, so that the EPS network releases the bearer dedicated.

ASR and ASA messages allow the PCRF entity to notify the AF entity that session on the EPS network is complete, for example as a result of the loss of coverage of the mobile.

DIAMETER messages exchanged over R9 interface between home PCRF (H-PCRF) and visited PCRF (V-PCRF) are identical to the messages exchanged over Gx and Rx interfaces.

Table 2.20 summarizes DIAMETER messages exchanged over the Gz interface between the PCEF entity and offline charging system (OFCS).

The same DIAMETER messages are exchanged over the Rf interface between, on the one hand, the entities of the IMS network, and, on the other hand, the OFCS entity.

Request	*Response*
Accounting-Request (ACR)	Accounting-Answer (ACA)

Table 2.20. *DIAMETER messages over Gz interface*

ACR and ACA messages enable the PCEF entity to inform the OFCS entity periodically on the consumption of resources in the case of post-paid service.

Table 2.21 summarizes DIAMETER messages exchanged over Gy interface between the PCEF entity and online charging system (OCS).

Request	*Response*
Credit-Control-Request (CCR)	Credit-Control-Answer (CCA)
Re-Auth-Request (RAR)	Re-Auth-Answer (RAA)
Abort-Session-Request (ASR)	Abort-Session-Answer (ASA)

Table 2.21. *DIAMETER messages over Gy interface*

The same DIAMETER messages are exchanged over the Ro interface between, on the one hand, media gateway control function (MGCF), IMS-GWF (gateway function) and TAS entities of the IMS network, and, on the other hand, the OCS entity.

CCR and CCA messages enable the PCEF entity to recover credit from OCS entity in the case of pre-paid service.

RAR and RAA messages allow the OCS entity to initialize a new authentication or new authorization.

ASR and ASA messages allow the OCS entity to end the current DIAMETER session.

3

Basic Procedures

3.1. Attachment

At the end of the connection procedure, the mobile starts the attachment procedure to the evolved packet system (EPS) which comprises the following steps:

– the mutual authentication between user equipment (UE) and mobility management entity (MME) corresponding to the AKA (Authentication and Key Agreement) mechanism;

– the security of non-access stratum (NAS) messages;

– the location of the mobile related to the tracking area identity (TAI) and to the E-UTRAN cell global identifier (ECGI);

– the establishment of a default bearer. In the case where the voice or video conversational call is supported by the EPS network, a default bearer (QCI = 5) is established for transporting the signaling exchanged between the mobile and the IP multimedia sub-system (IMS);

– the granting of a globally unique temporary identity (GUTI).

The UE mobile attachment procedure to the EPS network is described in Figure 3.1:

1) The attachment procedure is triggered by the UE entity when it sends to the eNB entity the EMM ATTACH REQUEST message containing the international mobile subscriber identity (IMSI) of the mobile.

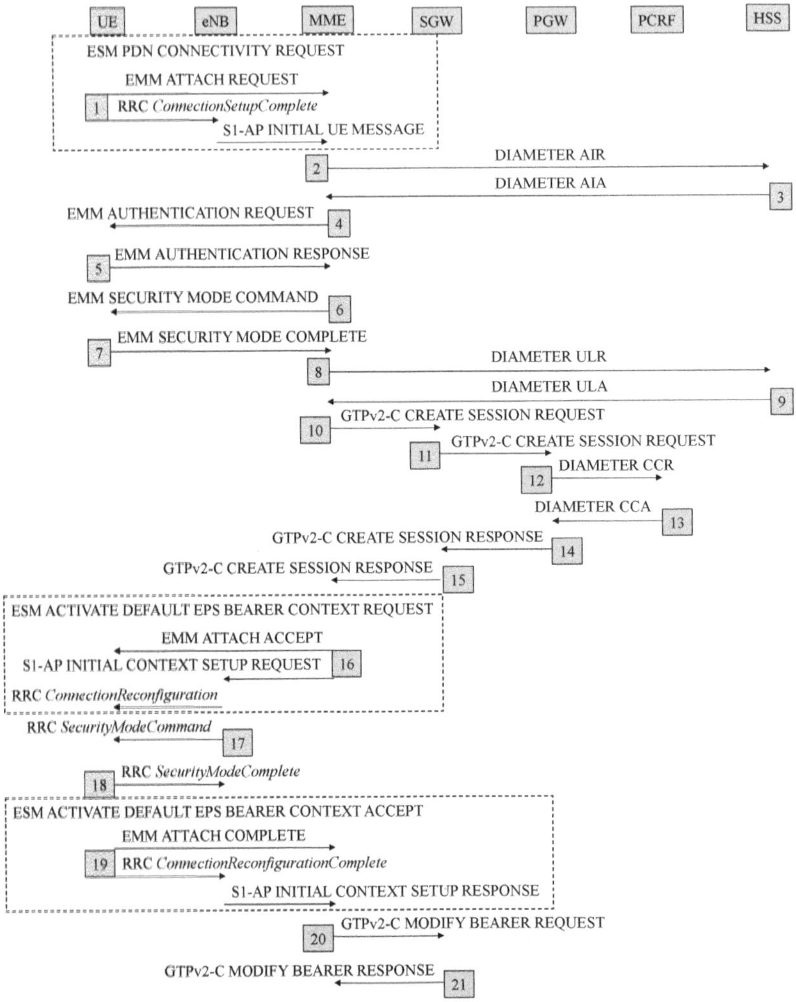

Figure 3.1. *Mobile attachment to EPS network*

The EMM ATTACH message carries the ESM PDN CONNECTIVITY REQUEST message.

The EMM ATTACH REQUEST message is carried by the RRC *ConnectionSetupComplete* message on the LTE-Uu radio interface and the S1-AP INITIAL UE MESSAGE message on the S1-MME interface.

The S1-AP UE INITIAL MESSAGE message contains the identities of TAI and ECGI.

2) Upon receipt of the EMM ATTACH REQUEST message, the MME entity requests the home subscriber server (HSS) entity for the cryptographic data of the mobile in the DIAMETER AIR (Authentication-Information-Request) message.

3) The HSS entity generates cryptographic data using a random number RAND and the key Ki of the mobile and sends them to the MME entity in the DIAMETER AIA (Authentication-Information-Answer) message.

The cryptographic data contain the random number (RAND), the mobile authentication code (RES), the network authentication code (AUTN) and the K_{ASME} key.

The MME entity derives the K_{ASME} key to generate the CK_{NAS}, IK_{NAS} and K_{eNB} keys:

– the CK_{NAS} key is used to encrypt the NAS message;

– the IK_{NAS} key is used to control the integrity of the NAS message;

– the K_{eNB} key is transferred to the eNB entity.

4) The MME entity transmits the EMM AUTHENTICATION REQUEST message containing the random number (RAND) and the network authentication code (AUTN) to the mobile.

The mobile calculates locally from its secret Ki stored in the universal services identity module (USIM) of its universal integrated circuit card (UICC) and the RAND random number received, the key K_{ASME}, its authentication code (RES) and that of the network (AUTN) which it compares to the value received. If the two values are identical, the network is authenticated.

5) The mobile responds to the MME entity with the EMM AUTHENTICATION RESPONSE message containing the RES authentication code.

The MME entity compares the RES values received from the mobile and HSS entity. If the values are the same, the mobile is authenticated.

6) The NAS signaling security setting is enabled by the MME entity sending the EMM SECURITY MODE COMMAND message controlled with the integrity key IK_{NAS}. This message contains algorithms to derive the K_{ASME} key.

The mobile derives the K_{ASME} key to generate the CK_{NAS}, IK_{NAS} and K_{eNB} keys and checks the integrity of the EMM SECURITY MODE COMMAND message.

7) The mobile responds with the EMM SECURITY MODE COMPLETE message encrypted with the CK_{NAS} key and controlled with integrity key IK_{NAS}.

After the mutual authentication phase and the security of NAS messages, the MME entity registers the mobile with the HSS entity.

8) The MME entity sends the DIAMETER Update-Location-Request (ULR) message to the HSS entity to register the mobile and obtain its service profile.

The HSS entity records the identity of the MME entity which attached the mobile and the identity of the TAI location area.

9) The HSS entity responds to the MME entity with the DIAMETER Update-Location-Answer (ULA) message that contains the profile of the mobile:

– the authorized access point names (APN);

– the characteristics of quality of service (QOS) for each default bearer that must be established.

The MME entity selects the serving gateway (SGW) entity in its group and the PDN gateway (PGW) entity from a domain name server (DNS) resolution of the APN.

10) The MME entity sends the GTPv2-C CREATE SESSION REQUEST message to create a context at the SGW entity.

The GTPv2-C CREATE SESSION REQUEST message contains the IP address of the PGW entity, the APN and the default bearer profile.

11) The SGW entity sends the GTPv2-C CREATE SESSION REQUEST message to create a context at the PGW entity.

The GTPv2-C CREATE SESSION REQUEST message contains the tunnel endpoint identifier (TEID) that the PGW entity will use at the GPRS tunneling protocol user plane (GTP-U) level for the S5 bearer.

12) The PGW entity sends to the policy and charging rules function (PCRF) the DIAMETER Credit-Control-Request (CCR) message for authorization to open the default bearer.

The PCRF entity compares the profile of the mobile with the rules defined for the network and stored in the subscription profile repository (SPR) database.

13) The PCRF entity responds to the PGW entity by a DIAMETER Credit-Control-Answer (CCA) message containing the rules to be applied to the default bearer (filter parameters and charging mode).

14) The PGW entity responds to the SGW entity by a GTPv2-C CREATE SESSION RESPONSE message which contains the following information:

– the TEID identifier that the SGW entity will use at the GTP-U protocol level for the S5 bearer;

– the mobile configuration (IP address of the mobile, IP address of its DNS server).

15) The SGW entity responds to the MME entity with a GTPv2-C CREATE SESSION RESPONSE which contains the following information:

– the TEID identifier that the eNB entity will use at the GTP-U protocol level for the S1 bearer;

– the mobile configuration.

16) The MME entity responds to the mobile with the EMM ATTACH ACCEPT message containing the following information:

– the mobile configuration;

– its GUTI temporary identity.

The EMM ATTACH ACCEPT message carries the ESM ACTIVATE DEFAULT EPS BEARER CONTEXT REQUEST message.

The EMM ATTACH ACCEPT message is carried by the S1-AP INITIAL CONTEXT SETUP REQUEST message on the S1-MME interface and by the RRC *ConnectionReconfiguration* message on the LTE-Uu radio interface.

The S1-AP INITIAL CONTEXT SETUP REQUEST message allows for the creation of the mobile context at the eNB entity level and contains the following information:

– the TEID identifier assigned by the SGW entity;

– the QoS parameters;

– the K_{eNB} key derived from the K_{ASME} key.

The eNB entity derives the key for creating the following keys:

– the K_{RRCenc} key for RRC message encryption;

– the K_{RRCint} key for RRC message integrity control;

– the K_{UPenc} key for traffic encryption.

The RRC *ConnectionReconfiguration* message initializes mounting of the data radio bearer (DRB).

17) The eNB entity requests the mobile to secure the radio interface with the RRC *SecurityModeCommand* message which is controlled with the integrity key K_{RRCint} and contains algorithms that allow the mobile to derive the K_{eNB} key.

The mobile derives the K_{eNB} key to generate the keys K_{RRCenc}, K_{RRCint} and K_{UPenc}, and checks the integrity of the RRC *SecurityModeCommand* message.

18) The mobile confirms the establishment of the keys to the eNB entity with the RRC *SecurityModeComplete* message which is controlled with the integrity key K_{RRCint}.

The messages of the last two steps are interposed between the reception of the S1-AP INITIAL CONTEXT SETUP REQUEST message and the transmission of the RRC *ConnectionReconfiguration* message by the eNB entity.

19) The mobile confirms receipt of the EMM ATTACH REQUEST message by sending the EMM ATTACH COMPLETE message.

The EMM ATTACH COMPLETE message carries the ESM ACTIVATE DEFAULT EPS BEARER CONTEXT ACCEPT message.

The EMM ATTACH COMPLETE message is carried by the RRC *ConnectionReconfigurationComplete* message on the LTE-Uu radio interface and by the S1-AP INITIAL CONTEXT SETUP RESPONSE message on the S1-MME interface.

The S1-AP INITIAL CONTEXT SETUP RESPONSE message contains the TEID identifier which the SGW entity will use at the GTP-U protocol level for the S1 bearer.

The RRC *ConnectionReconfigurationComplete* message acknowledges the reception of the RRC *ConnectionReconfiguration* message.

20) The MME entity transfers the TEID identifier received from the eNB entity to the SGW entity in the GTPv2-C MODIFY BEARER REQUEST message.

21) The SGW entity acknowledges the message received by the GTPv2-C MODIFY BEARER RESPONSE message.

3.2. Registration

The registration process to the IMS network includes the following steps:

– mutual authentication between the mobile and the serving call session control function (S-CSCF);

– establishment of an IPSec security association between the mobile and the P-CSCF entity (proxy-CSCF);

– registration by the S-CSCF entity of the correspondence between the IP address of the mobile and its uniform resource identifier (URI);

– subscription by the mobile and the P-CSCF entity for mobile registration events;

– notification by the S-CSCF entity of the events concerning the registration of the mobile.

The registration process to the IMS network is shown in Figure 3.2.

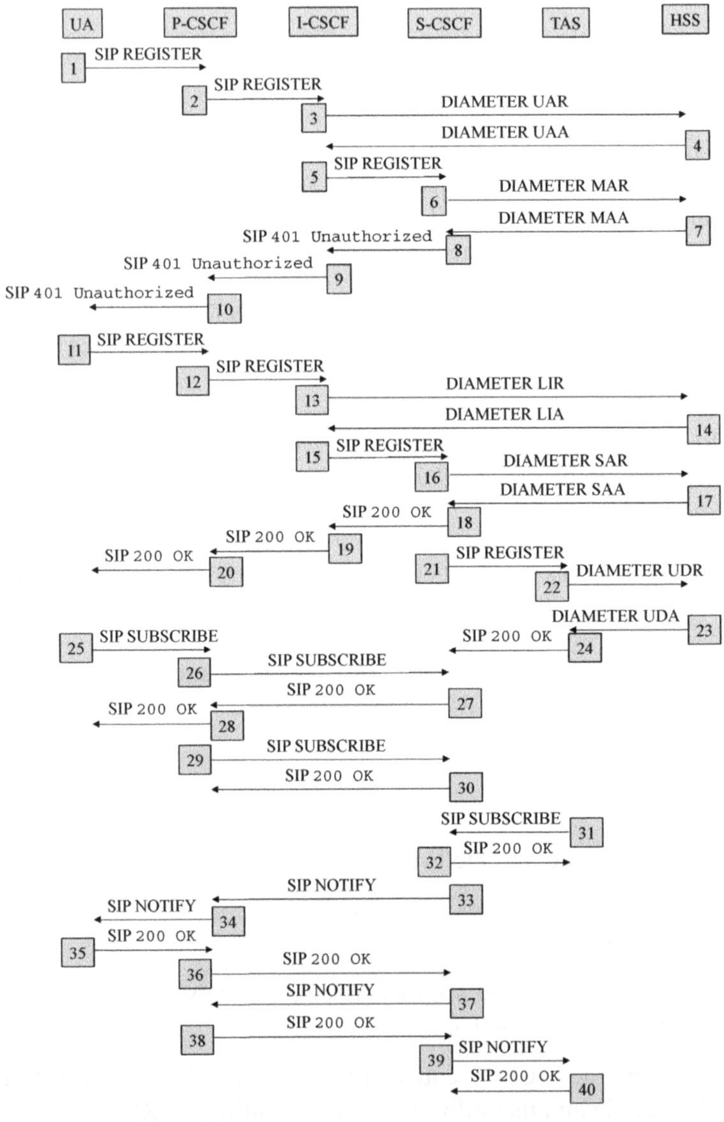

Figure 3.2. *Mobile registration to IMS network*

1) Alice's user agent (UA) sends to the P-CSCF entity an initial REGISTER request containing her private identity (alice_private@homeA.net) in the parameter username in the header Authorization.

The header To contains the public identity of Alice's UA entity (sip: alice@homeA.net).

The header Contact contains the IP address of Alice's UA entity.

The header Security-Client contains the parameters defining the security association IPSec established with the P-CSCF entity.

The headers Require and Proxy-Require contain the value sec-agree, indicating that the security mechanism based on IPSec is required.

The header Proxy-Require is addressed to the P-CSCF entity. The header Require is addressed to the S-CSCF entity in case the request is transmitted directly to it.

```
REGISTER sip:ims.mnc01.mcc208.3gppnetwork.org SIP/2.0
...
To: <sip:alice@homeA.net>
Contact: <sip:192.0.2.101>

Authorization: Digest
username="alice_private@homeA.net", realm="
ims.mnc01.mcc208.3gppnetwork.org ", nonce="",
uri="sip:ims.mnc01.mcc208.3gppnetwork.org", response=""
Security-Client: ipsec-3gpp; alg=hmac-sha-1-96; spi-
c=23456789; spi-s=12345678; port-c=2468; port-s=1357
Require: sec-agree
Proxy-Require: sec-agree
...
```

2) The P-CSCF entity transfers the REGISTER message to the I-CSCF (Interrogating-CSCF) entity, including its identity in the header Path.

The P-CSCF entity can find the IP address of the I-CSCF entity by doing a domain name system (DNS) resolution on the basis of the domain name of Alice's UA entity.

The P-CSCF entity provides the type of mobile network and the identity of the cell in the header `P-Access-Network-Info`. This information is provided by the PCRF entity.

The P-CSCF inserts the header `Require` containing the value `path` to ensure that the S-CSCF (Serving-CSCF) entity will take account of the header `Path`. If the S-CSCF entity does not support this header, it sends back a response `420 Bad extension`.

The P-CSCF entity declares in the parameter `integrity-protected` of the header `Authorization` that security has not been established with the UA entity.

The P-CSCF entity removes the headers `Require` and `Proxy-Require` which contained the `sec-agree` because IPSec security will be implemented by the P-CSCF entity.

```
REGISTER sip:ims.mnc01.mcc208.3gppnetwork.org SIP/2.0
...
Path: <sip:pcscf@homeA.net;lr>
Require: path
Authorization: Digest username="alice_private@homeA.net",
realm=" ims.mnc01.mcc208.3gppnetwork.org", nonce="",
uri="sip:ims.mnc01.mcc208.3gppnetwork.org ", response="",
integrity-protected="no"
...
```

3) I-CSCF entity transmits to the HSS entity the DIAMETER UAR (User-Authorization-Request) message to retrieve the list of S-CSCF entities that can be assigned to the UA entity.

4) I-CSCF entity performs the selection of an S-CSCF entity to which it forwards the REGISTER request from the list of S-CSCF entities received in the DIAMETER User-Authorization-Answer (UAA) message.

5) The I-CSCF entity replaces the initial URI (`sip:ims.mnc01.mcc208.3gppnetwork.org`) with that of the S-CSCF (`sip:scscf.homeA.net`)\ and sends the REGISTER request to the S-CSCF entity selected.

```
REGISTER sip:scscf.homeA.net SIP/2.0
...
```

6) The S-CSCF entity transmits to the HSS entity the DIAMETER MAR (Multimedia-Authentication-Request) message to recover the authentication data of the mobile.

7) The S-CSCF entity receives from the HSS entity the DIAMETER MAA (Multimedia-Authentication-Answer) message containing the random number (RAND), the mobile authentication code (RES), the network authentication code (AUTN), the integrity key (IK) and the cipher key (CK) for establishing the IPSec security association.

At this stage, the identity of the S-CSCF entity is registered in the HSS entity, and that of the P-CSCF entity in the S-CSCF entity.

8) The S-CSCF entity responds with a `401 Unauthorized` message containing the random number (RAND), the network authentication code (AUTN), the integrity key (IK) and the cipher key (CK).

The SIP `401 Unauthorized` response is transmitted to the I-CSCF entity whose identity is contained in the first `Via` header.

```
SIP/2.0 401 Unauthorized
...
WWW-Authenticate:
Digest realm="ims.mnc01.mcc208.3gppnetwork.org ",
nonce=base64(RAND + AUTN + server specific data),
algorithm=AKAv1-MD5,
ik="00112233445566778899aabbccddeeff",
ck="ffeeddccbbaa11223344556677889900"
...
```

9) After removing the `Via` header containing its identity, the I-CSCF entity forwards the response to the P-CSCF entity whose identity is contained in the first `Via` header.

10) After removing the `Via` header containing its identity and the keys IK and CK from the header `WWW-Authenticate`, the P-CSCF entity transfers the response to the Alice's UA entity (or Bob's) whose identity is contained in the first `Via` header.

In the header `Security-Server`, the P-CSCF entity indicates the parameters of the security association IPSec established with Alice's UA entity.

```
SIP/2.0 401 Unauthorized
...
WWW-Authenticate:
Digest realm="ims.mnc01.mcc208.3gppnetwork.org ",
nonce=base64(RAND + AUTN + server specific data),
algorithm=AKAv1-MD5
Security-Server: ipsec-3gpp; q=0.1; alg=hmac-sha-1-96;
spi-c=98765432; spi-s=87654321; port-c=8642; port-s=7531
...
```

Upon receiving a `401 Unauthorized` message, the mobile retrieves the key Ki from the IMS services identity module (ISIM) contained in the UICC card, calculates locally from the random number (RAND) received, the authentication code AUTN and RES and the keys IK and CK.

11) If the authentication code calculated, AU is identical to that received, the IMS network is authenticated and the Alice's UA entity (or Bob's) sends to the P-CSCF entity a second SIP REGISTER request comprising its identity private, the authentication code RES answer in parameter `response` of the `Authorization` header.

The REGISTER request, in the header `Security-Verify`, includes the security data received in the header `Security-Server` in the response `401 Unauthorized`.

```
REGISTER sip:ims.mnc01.mcc208.3gppnetwork.org SIP/2.0
...
Authorization: Digest
username="alice_private@homeA.net",
realm="ims.mnc01.mcc208.3gppnetwork.org",
nonce=base64(RAND + AUTN + server specific data),
algorithm=AKAv1-MD5,
uri="sip:ims.mnc01.mcc208.3gppnetwork.org",
response="6629fae49393a05397450978507c4ef1"
Security-Client: ipsec-3gpp; alg=hmac-sha-1-96; spi-
c=23456789; spi-s=12345678; port-c=2468; port-s=1357
Security-Verify: ipsec-3gpp; q=0.1; alg=hmac-sha-1-96;
spi-c=98765432; spi-s=87654321; port-c=8642; port-s=7531
...
```

12) The P-CSCF entity transfers the message REGISTER to the I-CSCF entity.

13) I-CSCF entity transmits to the HSS entity the DIAMETER LIR (Location-Information-Request) message to retrieve the IP address of S-CSCF entity.

14) The HSS entity provides the IP address of the S-CSCF entity in the DIAMETER LIA (Location-Information-Answer) message.

15) I-CSCF entity sends to the S-CSCF the REGISTER request.

The S-CSCF entity compares the value of the RES received from the UA entity with that of the RES received from the HSS entity. If the two values are identical, the mobile is authenticated.

16) The S-CSCF entity sends to the HSS entity the DIAMETER SAR (Server-Assignment-Request) message to retrieve the mobile profile.

17) The HSS entity transmits the profile of the mobile in a DIAMETER SAA (Server-Assignment-Answer) message.

The mobile profile contains the logic to invoke the telephony application server (TAS) that provides supplementary telephone services and authorized media.

The S-CSCF entity responds to the mobile with a SIP `200 OK` message including its identity in the `Service Route` header.

The registration is effective for a duration indicated in the parameter `expires` in the header `Contact`.

The mobile has to renew its registration before expiration of that time, by way of the same procedure as for initial registration.

The S-CSCF entity indicates, in the header `P-Associated-URI`, the identities registered implicitly, in addition to that indicated in the header `to`.

```
SIP/2.0 200 OK
...
Contact: <sip:192.0.2.101>;expires=600000
P-Associated-URI: <sip:+1-212-555-1111@homeA.net;
user=phone>
...
```

18), 19), 20) The SIP `200 OK` response follows the reverse path taken by the REGISTER request, through the `Via` headers.

21) The S-CSCF entity registers the mobile with the TAS entity.

22) The TAS entity sends to the HSS entity the DIAMETER UDR (User-Data-Request) message to retrieve the mobile service data.

23) The answer is provided by the DIAMETER UDA (User-Data-Answer) message.

24) The TAS entity responds with `200 OK` message to the SIP REGISTER request.

25) Alice's UA entity (or Bob's) generates a SUBSCRIBE request to receive notifications for registration.

The identity of the S-CSCF entity of the domain (`homeA.net`), contained in the header `Route`, is learned during the registration of Alice's UA entity, information from the header `Service Route`.

If Alice's UA entity has multiple identities, it indicates in the header `P-Preferred-Identity` which among them is the preferred one (`sip:alice@homeA.net`).

Alice's UA entity defines in the header `Event` the type of event which it wishes to subscribe (value `reg` for registration).

Alice's UA entity publishes the duration of the subscription in the header `Expires`. The value must not be greater than that indicated during registration in the response `200 OK` of the REGISTER request.

Alice's UA entity defines the format of the notifications of registration events in the header `Accept` (`application/reginfo+xml`).

26) The P-CSCF entity removes the headers `Route` containing its own identity and replaces the header `P-Preferred-Identity` with the header `P-Asserted-Identity` containing Alice's asserted URI identity.

The P-CSCF adds the headers `Via` and `Record Route` containing its own identity. The header `Record Route` enables us to construct the route taken by subsequent requests.

The S-CSCF authorizes the subscription with the response `200 OK` because it trusts the content of the header `P-Asserted-Identity` inserted by the P-CSCF and because Alice has subscribed to the registration event service.

27), 28) The response `200 OK` is routed through the addresses contained in the headers `Via`.

29) Receipt of the message `200 OK` triggers a subscription on the part of the P-CSCF entity to discover the state of registration of Alice's UA entity.

30) The S-CSCF entity responds with `200 OK`.

31) The `200 OK` response to the REGISTER message triggers the SUBSCRIBE request from the TAS entity to have knowledge of the registration status of Alice's UA entity (or Bob's).

32) The S-CSCF entity responds with `200 OK`.

33), 34) The S-CSCF entity notifies Alice's UA entity of the registration of its identities by sending a NOTIFY request.

The S-CSCF reports the state of the subscription in the header `Subscription-State` (value active) and its duration in the parameter `expires`.

The notification of registration elements is produced in an XML (eXtensible markup language) document of type `reginfo` attached to the SIP (session initiation protocol) message.

35), 36) The `200 OK` response of the mobile is routed from addresses contained in the `Via` headers.

37) The S-CSCF entity notifies to the P-CSCF entity the registration status of Alice's UA entity (or Bob's) by sending a NOTIFY request.

38) The P-CSCF entity responds with `200 OK`.

39) The S-CSCF entity notifies to the TAS entity the registration status of Alice's UA entity (or Bob's) by sending a NOTIFY request.

40) The TAS entity responds with `200 OK`.

3.3. Deregistration

The deregistration process to the IMS network is described in Figure 3.3.

1), 2) The deregistration phase starts when the Alice's UA entity (or Bob's) transmits a REGISTER message which parameter `expires` of the header `Contact` has a zero value.

3) The I-CSCF entity transmits to the HSS entity the DIAMETER LIR message to retrieve the IP address of S-CSCF.

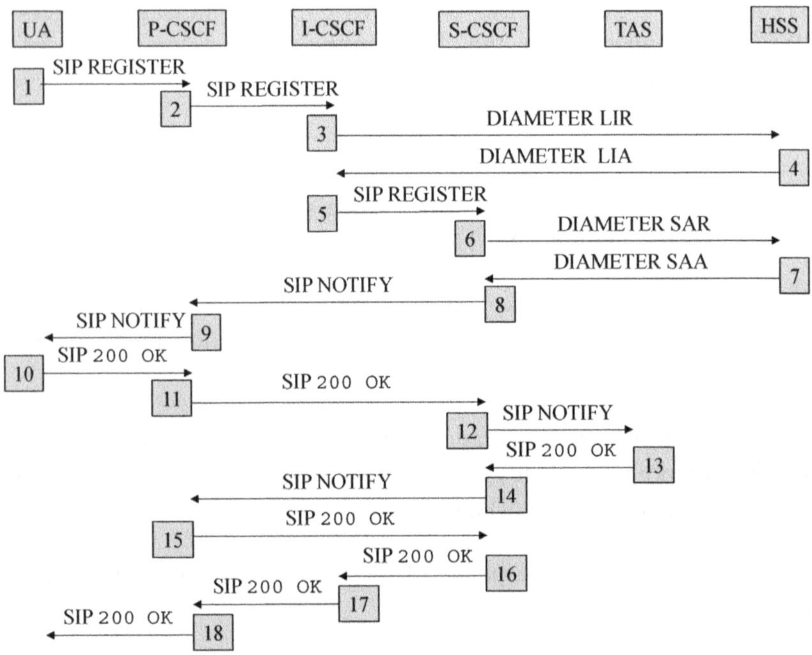

Figure 3.3. *Mobile deregistration to IMS network*

4) The HSS entity provides the IP address of the S-CSCF entity in the DIAMETER LIA message.

5) The I-CSCF entity sends to the S-CSCF entity the REGISTER request.

6) The S-CSCF entity transmits to the HSS entity the DIAMETER SAR message informing about deregistration of the mobile.

The HSS entity still retains the identity of the S-CSCF entity to allow the use of services when the mobile is not registered, such as call forwarding to voicemail.

7) The HSS entity transmits the DIAMETER SAA message to acknowledge the previous request.

8), 9) The S-CSCF entity informs Alice's UA entity (or Bob's) of its deregistration in a NOTIFY message.

10), 11) The `200 OK` message is a response to the NOTIFY request.

12) The S-CSCF entity informs the TAS entity about the deregistration of Alice's UA entity in a NOTIFY message.

13) The `200 OK` message is a response to the NOTIFY request.

14) The S-CSCF entity informs the P-CSCF entity about the deregistration of Alice's UA entity in a NOTIFY message.

15) The `200 OK` message is a response to the NOTIFY request.

16), 17), 18) The 200 OK message is a response to the REGISTER request.

3.4. Detachment

The detachment procedure to the EPS network is described in Figure 3.4.

1) Upon receipt of `200 OK` response to the REGISTER request for deregistration, Alice's UA entity (or Bob's) starts the detachment phase and passes to the MME entity the EMM DETACH REQUEST message.

EMM ATTACH REQUEST message carries ESM PDN DISCONNECT REQUEST message.

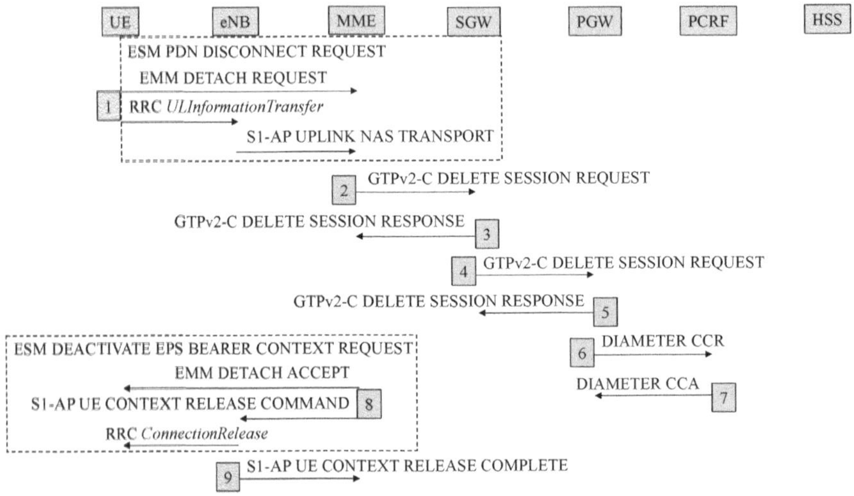

Figure 3.4 *Mobile detachment to EPS network*

EMM ATTACH REQUEST message is carried by RRC *ULInformation Transfer* message on the radio interface LTE-Uu and S1-AP UPLINK NAS TRANSPORT message on the S1-MME interface.

2) The MME entity sends to the SGW entity the GTPv2-C DELETE SESSION REQUEST message to deactivate the default bearer used for telephone signaling.

3) The SGW entity responds with the GTPv2-C DELETE SESSION RESPONSE message.

4) The SGW entity sends to the PGW entity the GTPv2-C DELETE SESSION REQUEST message to deactivate the default bearer.

5) The PGW entity responds with the GTPv2-C DELETE SESSION RESPONSE message.

6) The PGW entity sends to the PCRF entity the DIAMETER CCR message to inform about the deactivation of the default bearer.

7) The PCRF entity responds to the PGW entity with the DIAMETER CCA message.

8) The MME entity confirms the detachment to Alice's UA entity in the EMM DETACH ACCEPT message.

EMM DETACH ACCEPT message carries ESM DEACTIVATE EPS BEARER CONTEXT REQUEST message.

EMM DETACH ACCEPT message is carried by RRC *Connection Release* message on the radio interface LTE-Uu and S1-AP UE CONTEXT RELEASE COMMAND message on the S1-MME interface.

9) The eNB entity confirms the context deactivation in S1-AP EU CONTEXT RELEASE COMPLETE message.

3.5. Establishment of VoLTE session

The establishment of the VoLTE session includes the following operations:

– the negotiation of the characteristics of real-time transport protocol (RTP) streams via session description protocol (SDP);

– the establishment of dedicated bearer to the RTP stream.

3.5.1. *Originating side*

The procedure for establishing the VoLTE session on the outgoing call is described in Figure 3.5.

1) Alice's UA entity `generates` an initial INVITE request sent to Bob's URI (`sip:bob@homeB.net`).

Alice's UA entity specifies in the header `Require` that Bob's UA entity must support the precondition of resource reservation before activating the ringing of the telephone.

Alice's UA entity indicates in the header `Supported` that it supports the acknowledgement of provisional responses of `1xx` type (value `100rel`).

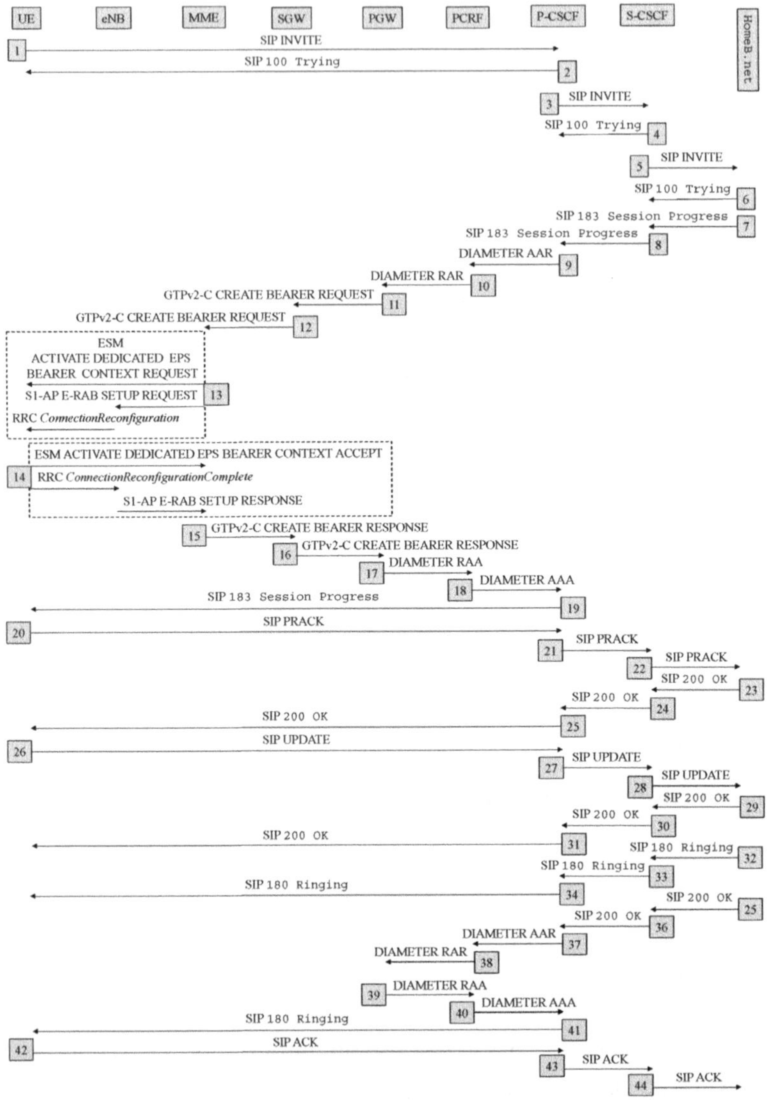

Figure 3.5. *Establishment of VoLTE session: originating side*

Alice's UA entity indicates in the header `Allow` the different supported methods (INVITE, ACK, CANCEL, BYE, PRACK, UPDATE, REFER and MESSAGE).

Alice's UA entity sends an SDP offer in the initial INVITE request to Bob's UA entity. The offer lists the media (audio) that Alice wishes to use for that session and lists the different codecs supported as well.

Alice's UA entity indicates in the SDP offer (`a=curr:qos local none`, `a=curr:qos remote none`) that the resource reservation has not been established for the local and remote UA entities.

Alice's UA entity indicates in the SDP offer (`a=des:qos mandatory local sendrecv`) that the precondition of resource reservation is mandatory for the local UA entity.

Alice's UA entity indicates in the SDP offer (`a=des:qos none remote sendrecv`) that the precondition of resource reservation is unknown for the remote UA entity.

```
INVITE sip:bob@homeB.net SIP/2.0
...
Require: precondition, sec-agree
Proxy-Require: sec-agree
Supported: 100rel
Allow: INVITE, ACK, CANCEL, BYE, PRACK, UPDATE, REFER,
MESSAGE
...
Content-Type: application/sdp
Content-Length: (...)
...
a=curr:qos local none
a=curr:qos remote none
a=des:qos mandatory local sendrecv
a=des:qos none remote sendrecv
...
```

2) The P-CSCF entity responds to the mobile with `100 Trying` message that allows for blocking the retransmission timer of the INVITE request.

3) The P-CSCF entity removes the header `Route` containing its own identity.

The P-CSCF entity replaces the header `P-Preferred-Identity` with the header `P-Asserted-Identity` containing the asserted URI of Alice.

The P-CSCF entity adds the headers `Via` and `Record Route` containing its own identity. The header `Record Route` enables us to construct the route to be taken by subsequent requests.

The P-CSCF entity forwards the request to the S-CSCF entity whose identity is contained in the header `Route`.

4) The S-CSCF entity responds to P-CSCF entity with `100 Trying` message that allows for blocking the retransmission timer of the INVITE request.

5) The S-CSCF entity removes the header `Route` containing its own identity.

The S-CSCF entity adds the headers `Via` and `Record Route` containing its own identity.

The HSS entity has provided, during registration, the S-CSCF entity with the user profile, containing the types of media authorized by the service offered.

The S-CSCF entity examines the information contained in the SDP message carried by the INVITE request. If the S-CSCF finds that these do not conform to the service profile, it sends to Alice's UA entity a negative response `488 Not Acceptable Here`.

The S-CSCF entity retrieves the destination domain name (`homeB.net`) from Bob's URI and forwards the INVITE request to the entity in charge of the interconnection with the domain (`homeB.net`).

6) The `homeB.net` domain responds to S-CSCF entity with `100 Trying` message that allows for blocking the retransmission timer of the INVITE request.

7) In the response `183 Session Progress`, Bob's UA entity gives an SDP response in which it chooses a type of codec from those proposed by Alice's UA entity.

The `183 Session Progress` response contains in the header `Record Route` IP addresses of entities that the subsequent requests must pass through.

The `183 Session Progress` response also indicates that a resource reservation is also required of Bob's side before establishing a session.

```
...
a=curr:qos local none
a=curr:qos remote none
a=des:qos mandatory local sendrecv
a=des:qos mandatory remote sendrecv
...
```

8) The S-CSCF entity forwards the `183 Session Progress` response to the P-CSCF entity. The identity of the P-CSCF entity assigned to Bob's UA entity was learned during the registration (information contained in the `Path` header).

The establishment of a dedicated bearer, assigned to voice, is coupled with the establishment of a default bearer and assigned to telephone signaling.

This dedicated bearer is coupled with the default bearer assigned to telephone signaling, in that the bearer terminations are the same for both types of bearer.

The establishment of the dedicated bearer is triggered by the P-CSCF entity from the analysis of telephone signaling exchanged between the terminals that want to establish a telephone call.

9) The P-CSCF entity sends to the PCRF entity the DIAMETER AAR (authenticate and authorize request) message to verify that the media settings are authorized by the EPS network and to trigger the establishment of dedicated bearer.

10) The PCRF entity consults its SPR database to retrieve the level of QoS to apply (QCI = 1) and transmits the DIAMETER RAR (Re-Auth-Request) message to the PGW entity containing the characteristics of the stream to establish.

11) The PGW entity sends to the SGW entity the GTPv2-C CREATE BEARER REQUEST message containing the TEID identifier of the S5 bearer that the SGW entity will need to use in the GTP-U header when sending traffic to the PGW entity.

12) The SGW entity sends to the MME entity the GTPv2-C CREATE BEARER REQUEST message containing the TEID identifier of the S1 bearer that the eNB entity will need to use in the GTP-U header while sending traffic to the SGW entity.

13) The MME entity sends to the mobile the ESM ACTIVATE DEDICATED EPS BEARER CONTEXT REQUEST message carried by the S1-AP E-RAB SETUP REQUEST message on the S1-MME interface and by the RRC *ConnectionReconfiguration* message on the LTE-Uu interface.

The S1-AP E-RAB SETUP REQUEST message contains the TEID identifier that the MME entity has received from the SGW entity.

The RRC *ConnectionReconfiguration* message contains the logical channel identifier (LCID) of the dedicated bearer.

14) The mobile responds to the MME entity by the ESM ACTIVATE DEDICATED EPS BEARER CONTEXT ACCEPT message carried by the RRC *ConnectionReconfigurationComplete* message on the LTE-Uu interface and by the S1-AP E-RAB SETUP RESPONSE message on the S1-MME interface.

The S1-AP E-RAB SETUP RESPONSE message contains the TEID identifier of the S1 bearer that the SGW entity will need to use in the GTP-U header when it sends traffic to the eNB entity.

15) The MME entity responds to the SGW entity by the GTPv2-C CREATE BEARER RESPONSE message containing the TEID identifier that the MME entity has received from the eNB entity.

16) The SGW entity responds to the PGW entity with the C-GTPv2 CREATE BEARER RESPONSE message containing the TEID identifier of the S5 bearer that PGW entity will need to use in the GTP-U header when it sends traffic to the SGW entity.

The dedicated bearer has been established at this stage. It will remain blocked by the PGW entity until Bob picks up the call.

17) The PGW entity responds to the PCRF entity with the DIAMETER RAA (Re-Auth Answer) message.

18) The PCRF entity responds to the P-CSCF entity with the DIAMETER AAA (Authenticate-Authorize-Answer) message.

19) The P-CSCF entity forwards the `183 Session Progress` response to the Alice's UA entity.

20), 21), 22) Alice's entity sends the subsequent PRACK request to acknowledge the provisional response `183 Session Progress`.

Alice's UA entity indicates in the `Route` headers the identities of CSCF entities processing the request, namely P / C-CSCF entities in the domains (`homeA.net`) and (`homeB.net`).

23), 24), 25) The `200 OK` Message is the response to the PRACK request.

26), 27), 28) When Alice's UA entity has confirmation of the resource reservation, it indicates this to Bob's UA entity in an SDP offer (`a=curr:qos local sendrecv`) contained in an UPDATE request.

```
...
a=curr:qos local sendrecv
a=curr:qos remote none
a=des:qos mandatory local sendrecv
a=des:qos mandatory remote sendrecv
...
```

29), 30), 31) When Bob's UA entity has confirmation of the resource reservation, it notifies Alice's UA entity in an SDP offer contained in the response `200 OK` to the UPDATE request.

```
...
a=curr:qos local sendrecv
a=curr:qos remote sendrecv
a=des:qos mandatory local sendrecv
a=des:qos mandatory remote sendrecv
...
```

32), 33), 34) When the resource reservations are effective at both ends, Bob's telephone can ring and a response `180 Ringing` is transmitted to Alice's UA entity, generating a ring back tone for her.

35), 36) When Bob picks up the phone, the `200 OK` response to the INVITE request is sent to the PCSCF entity.

37) The P-CSCF entity sends to the PCRF entity the DIAMETER AAR message to indicate that Bob took the call.

38) The PCRF entity transmits to the PGW entity the RAR DIAMETER message to unblock the dedicated bearer.

39) The PGW entity responds to the PCRF entity with the DIAMETER AAR message.

40) The PCRF entity responds to the P-CSCF entity with the DIAMETER AAA message.

41) The P-CSCF entity forwards the `200 OK` response to the INVITE request to Alice's UA entity, which causes the end of the ring back tone.

42), 43), 44) Alice's UA entity acknowledges `200 OK` response by the subsequent ACK request.

3.5.2. *Terminating side*

The procedure for establishing the VoLTE session on the incoming call is shown in Figure 3.6.

1) The I-CSCF entity of the domain `homeB.net` receives the INVITE request from the domain `homeA.net`.

2) The I-CSCF entity responds to the `homeA.net` domain with `100 Trying` message that allows for blocking the retransmission timer of the INVITE request.

3) The I-CSCF entity transmits to the HSS entity the DIAMETER LIR message to retrieve the IP address of S-CSCF which recorded Bob's UA entity.

4) The HSS entity the IP address of the S-CSCF entity in the DIAMETER LIA message.

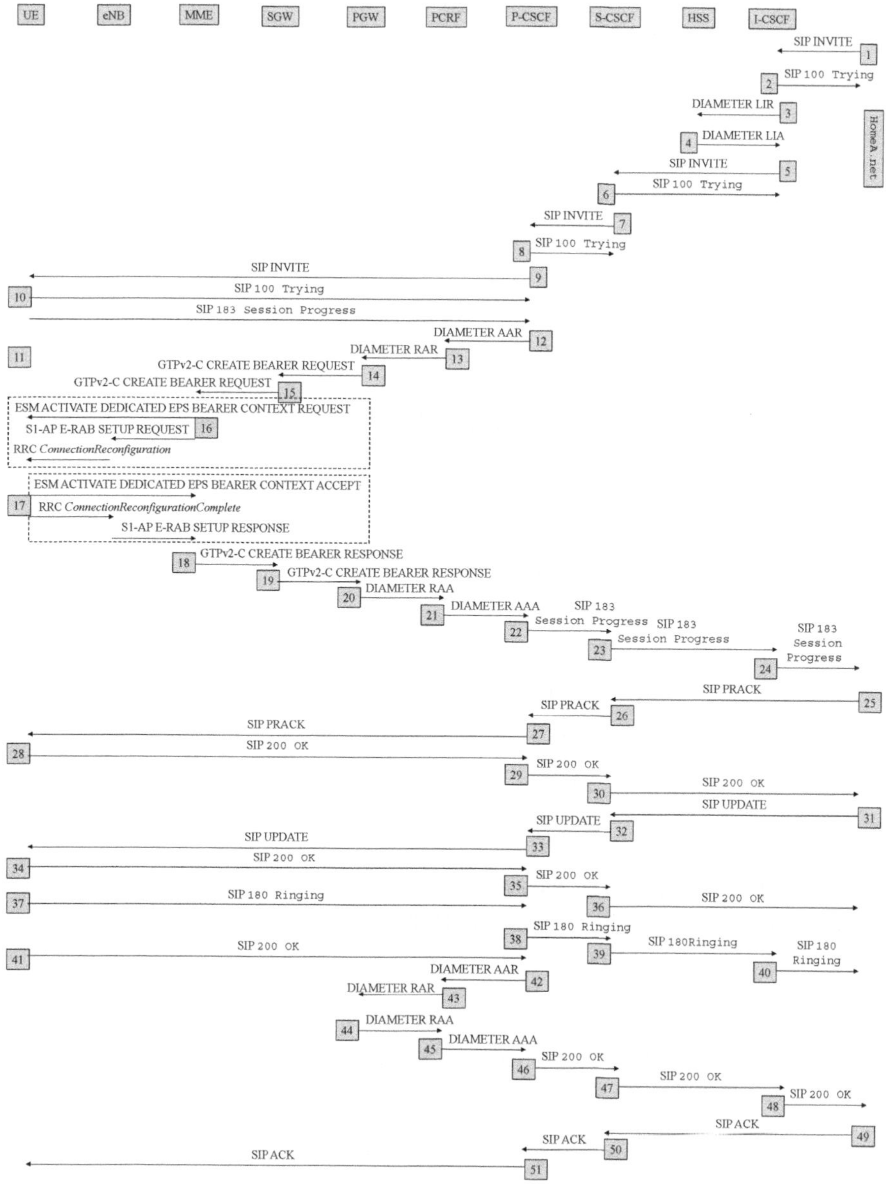

Figure 3.6. *Establishment of VoLTE session: terminating side*

5) The I-CSCF entity adds the header Via, does not add a Record Route and forwards the request to the S-CSCF entity. The subsequent requests, therefore, do not pass through the I-CSCF entity.

6) The S-CSCF entity responds to the I-CSCF entity with 100 Trying message that allows for blocking the retransmission timer of the INVITE request.

7) The S-CSCF entity verifies that the types of media proposed by Alice's UA entity correspond to the services offered to Bob.

The S-CSCF entity adds a header Record Route containing its own identity.

The S-CSCF entity modifies the initial request, replacing Bob's URI with his IP address. The link between the URI and the IP address was created at registration.

The S-CSCF entity adds Bob's URI to the header P-Called-Party-ID so that Bob knows which URI the INVITE request refers to.

The S-CSCF entity transfers the request to the P-CSCF entity. The IP address of the P-CSCF entity was learnt at registration (information contained in the header Path).

8) The P-CSCF entity responds to the S-CSCF entity with 100 Trying message that allows for blocking the retransmission timer of the INVITE request.

9) The P-CSCF entity adds a header Record Route containing its own identity and forwards the request to Bob's UA entity, whose IP address is included in the URI of the request.

10) Bob's UA entity entity responds to the P-CSCF entity with 100 Trying message that allows for blocking the retransmission timer of the INVITE request.

Bob's UA entity stores the different headers Record Route, which it will later use to route subsequent requests.

11) Bob's UA entity sends a response `183 Session Progress` to Alice's UA entity. The response is routed on the basis of the `Via` headers received from the request INVITE.

The response `183 Session Progress` contains the `Record Route` headers received from the request INVITE. This enables Alice's UA entity to retrieve the IP addresses of the CSCF entities that need to process the subsequent requests.

Bob's UA entity indicates the value `100rel` in the header `Require` to indicate that the response requires acknowledgement from Alice's UA entity. To correlate the acknowledgement with the response, Bob's UA entity inserts the header `RSeq`.

In the response `183 Session Progress`, Bob's UA entity gives an SDP response in which it chooses a type of codec from those proposed by Alice's UA entity.

Bob's UA entity indicates in the SDP message of the `183 Session Progress` response that resource reservation is also necessary on its part before establishing a session.

12) to 21) On receiving of the `183 Session Progress` response, the P-CSCF entity initiates the establishment of dedicated bearer. They are identical to messages 9 to 18 described in the preceding paragraph.

22), 23), 24) The `183 Session Progress` response is transmitted to the domain `homeA.net`.

25), 26), 27) Subsequent PRACK request is received from the domain `homeA.net`, indicating that Alice's UA entity has acknowledged receiving `183 Session Progress` response.

28), 29), 30) The `200 OK` message is the response to the PRACK request.

31), 32), 33) Subsequent UPDATE request is received from the domain `homeA.net`, indicating that the dedicated bearer on the Alice's side is established.

34), 35), 36) The `200 OK` message is the response to the UPDATE request, indicating that the dedicated bearer on the Bob's side is established.

37) to 40) The `180 Ringing` response to the initial INVITE request is sent to the domain `homeA.net`, indicating that the Bob's phone is ringing.

41) The `200 OK` response to the initial INVITE request indicates that Bob picks up his phone.

42) The P-CSCF entity sends the PCRF entity the DIAMETER AAR message to indicate that Bob took the call.

43) The PCRF entity transmits to the PGW entity the DIAMETER RAR message to unblock the dedicated bearer.

44) The PGW entity responds to the PCRF entity with the DIAMETER AAR message.

45) The PCRF entity responds to the P-CSCF entity with the DIAMETER AAA message.

46), 47), 48) The P-CSCF entity forwards the `200 OK` response to the INVITE request to the domain `homeA.net`.

49), 50), 51) The ACK request of the `200 OK` response is received from the domain `homeA.net`.

3.6. Termination of VoLTE session

The termination of the session can be triggered by UA, P-CSCF or IMS-GWF entity, using the BYE method.

Termination of the session can be initiated by either (Alice's or Bob's) UA entity when the communication has finished.

The P-CSCF entity can also end the session if the mobile is no longer within radio coverage. The PCRF entity sends this information to the P-CSCF entity by way of a DIAMETER ASR (Abort-Session-Request) message.

The IMS-GWF entity is an application server which controls the session in the case of pre-paid use. When the user's credit is exhausted, the IMS-GWF entity ends the session. Two BYE requests are needed to terminate the session: one request sent to Alice's UA entity and the second to Bob's UA entity.

3.6.1. *Initiated side*

The procedure for clearing the VoLTE session initiated by the UA entity, at the initiated side, is described in Figure 3.7.

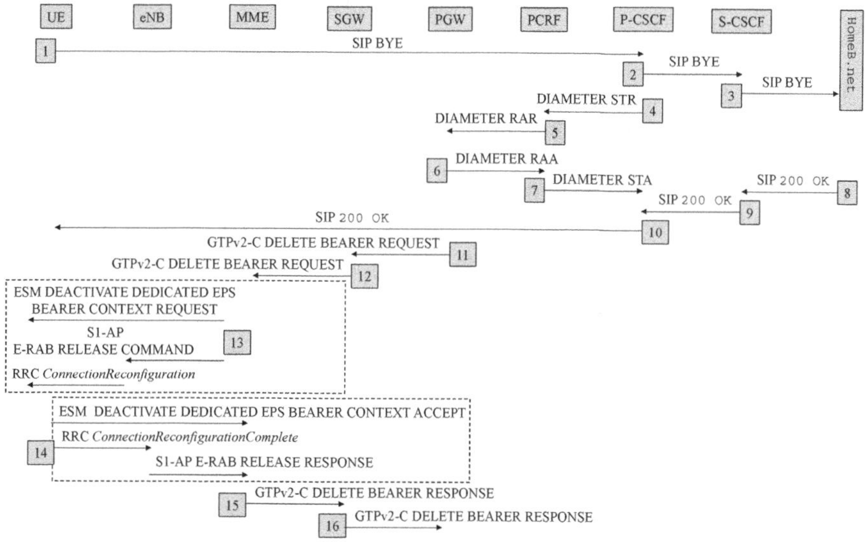

Figure 3.7. *Termination of VoLTE session: initiated side*

1), 2), 3) The release of the session is initiated by Alice's UA entity (or Bob's), which forwards the BYE request to the domain `homeB.net` (or `homeA.net`).

4) On receipt of the BYE request, the P-CSCF entity informs the PCRF entity for the release of the session in the DIAMETER STR (Session-Termination-Request) message.

5) The PCRF entity requests the PGW entity to release of dedicated bearer in the DIAMETER RAR message.

6) The PGW entity acknowledges the DIAMETER RAR request by the RAR DIAMETER RAA message.

7) The PCRF entity acknowledges the DIAMETER STR request by the DIAMETER STA (Answer-Session-Termination) message.

8), 9), 10) Alice's UA entity (or Bob's) receives the `200 OK` response to the BYE request from the domain `homeB.net` (or `homeA.net`).

11) Upon receipt of the DIAMETER RAR message, the PGW entity initiates the removal of dedicated bearer by the GTPv2 C-BEARER DELETE REQUEST message.

12) The SGW entity transmits the GTPv2 C-BEARER DELETE REQUEST message to continue the dedicated bearer release request.

13) The MME entity sends the ESM DEACTIVATE EPS BEARER CONTEXT REQUEST message to Alice's UA entity (or Bob's) to inform about the deactivation of the dedicated bearer.

The ESM DEACTIVATE EPS BEARER CONTEXT REQUEST message is carried by the RRC *ConnectionReconfiguration* message on the radio interface LTE-Uu and the S1-AP E-RAB RELEASE COMMAND message on the S1-MME interface.

14) The UA entity confirms the deactivation of the dedicated bearer in the ESM DEACTIVATE EPS BEARER CONTEXT ACCEPT message.

The ESM DEACTIVATE EPS BEARER CONTEXT ACCEPT message is carried by the RRC *ConnectionReconfigurationComplete* message on the radio interface LTE-Uu and the S1-AP E-RAB RELEASE RESPONSE message on the S1-MME interface.

15) The MME entity responds to the SGW entity with the GTPv2-C DELETE BEARER RESPONSE message that acknowledges the release request of the dedicated bearer.

16) The SGW entity responds to the PGW entity with the GTPv2-C DELETE BEARER RESPONSE message that acknowledges the release request of the dedicated bearer.

3.6.2. *Received side*

The procedure for clearing the VoLTE session initiated by the UA entity, at the received side, is described in Figure 3.8.

1), 2), 3) Bob's UA entity (or Alice's) receives the BYE request, from the domain homeA.net (or homeB.net), ending the session.

4) to 7) As described earlier, these messages are used to trigger the release of the dedicated bearer.

8), 9), 10) Bob's UA entity (or Alice's) sends the 200 OK response to the BYE request, to the domain homeA.net (or homeB.net).

11) to 16) As described earlier, these messages are used to remove the dedicated bearer.

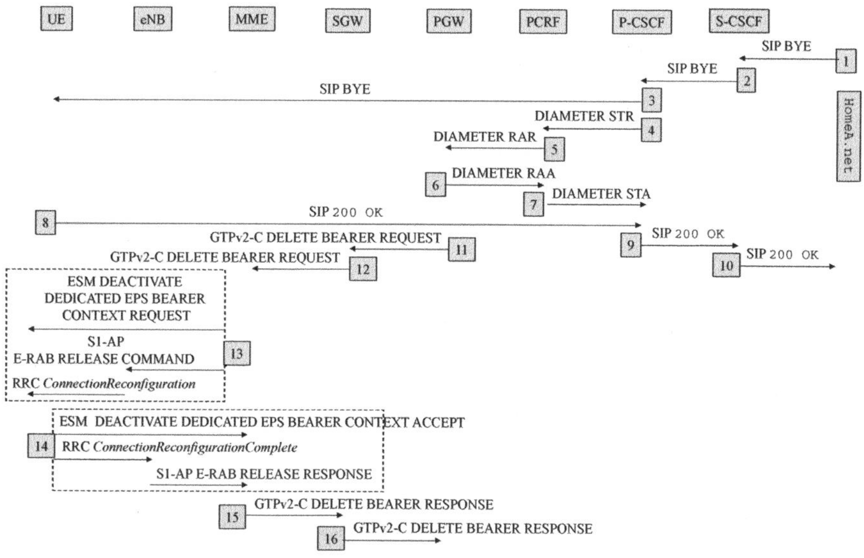

Figure 3.8. *Termination of VoLTE session: received side*

3.7. Establishment of ViLTE session

The procedure for establishing the ViLTE session is identical to that described for the VoLTE session when the SDP protocol associated with the initial INVITE request contains the negotiation of the two media, audio and video, with a proposal for codecs for both media.

Similarly, the 183 Session Progress response defines the codec selected among the proposals, for each media.

Upon receipt of the `183 Session Progress` response, the P-CSCF entity triggers to the PCRF entity to establish two dedicated bearers:

– one dedicated bearer (QCI = 1) for audio flow;

– one dedicated bearer (QCI=2) for video flow.

The ViLTE session can also be added to a VoLTE session.

The procedure of the addition of the video stream for the ViLTE session, at the initiated side, is described in Figure 3.9.

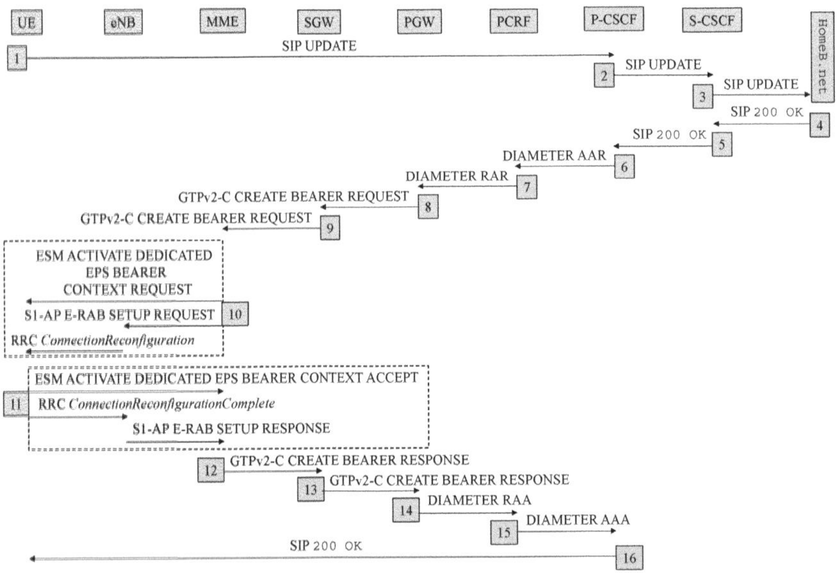

Figure 3.9. *Adding a video stream: initiated side*

1), 2), 3) Alice's UA entity (or Bob's) forwards the UPDATE request to the domain `homeB.net` (or `homeA.net`).

The UPDATE request can be replaced by the re-INVITE request and contains a new SDP offer on both audio and video streams.

The S-CSCF entity controls that Alice's UA entity (or Bob's) is authorized to establish a ViLTE session.

4), 5) The `200 OK` response to the UPDATE request contains the SDP response in which Bob's UA entity (or Alice's) retains a codec from those proposed by Alice's UA entity (or Bob's).

6) to 15) Upon receipt of the `200 OK` response, the P-CSCF entity initiates the establishment of dedicated bearer for the video stream.

16) The P-CSCF entity forwards the `200 OK` response to Alice's UA entity (or Bob's).

The procedure of the addition of the video stream for the ViLTE session, at the received side, is described in Figure 3.10.

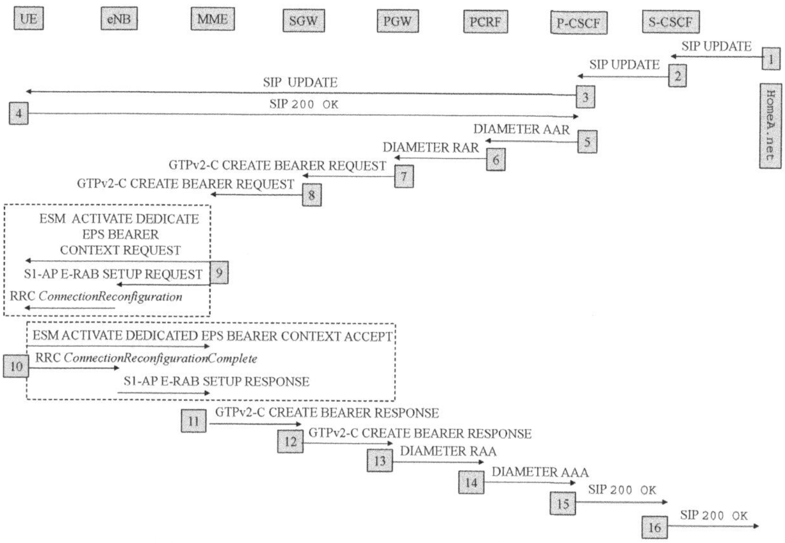

Figure 3.10. *Adding a video stream: received side*

1), 2), 3) The UPDATE request received from the domain `homeA.net` (or `homeB.net`) is transmitted to Bob's UA entity (or Alice's).

The S-CSCF entity controls that Bob's UA (or Alice's) is authorized to establish a ViLTE session.

4) Bob's UA entity (or Alice's) selects a codec for the video stream among the proposals received in the SDP offer associated with the UPDATE request.

The UA entity responds with `200 OK` message containing the selected codec.

5) to 14) Upon receipt of the `200 OK` response, the P-CSCF entity initiates the establishment of dedicated bearer for the video stream.

15), 16) The P-CSCF entity forwards the `200 OK` response to the domain `homeA.net` (or `homeB.net`).

3.8. Termination of ViLTE session

The procedure for clearing the ViLTE session is identical to that described for the VoLTE session if the two streams, audio and video, need to be released.

The release of the ViLTE session may be limited to the removal of the video stream, the audio stream being retained.

The procedure of the removal of the video stream, at the initiated side, is described in Figure 3.11.

1), 2), 3) Alice's UA entity (or Bob's) takes the initiative in removing the video flow by sending the UPDATE message to the domain `homeB.net` (or `homeA.net`).

The deletion of the video flow is indicated in the SDP message, which contains the port number of the video stream set to ZERO.

4) to 7) As described earlier, these messages are used to trigger the release of the dedicated bearer for the video stream.

8), 9), 10) Bob's UA entity (or Alice's) sends the `200 OK` response to the UPDATE request to the domain `homeA.net` (or `homeB.net`).

The deletion of the video flow is indicated in the SDP message, which contains the port number of the video stream set to ZERO.

11) to 16) As described earlier, these messages are used to remove the dedicated bearer for the video stream.

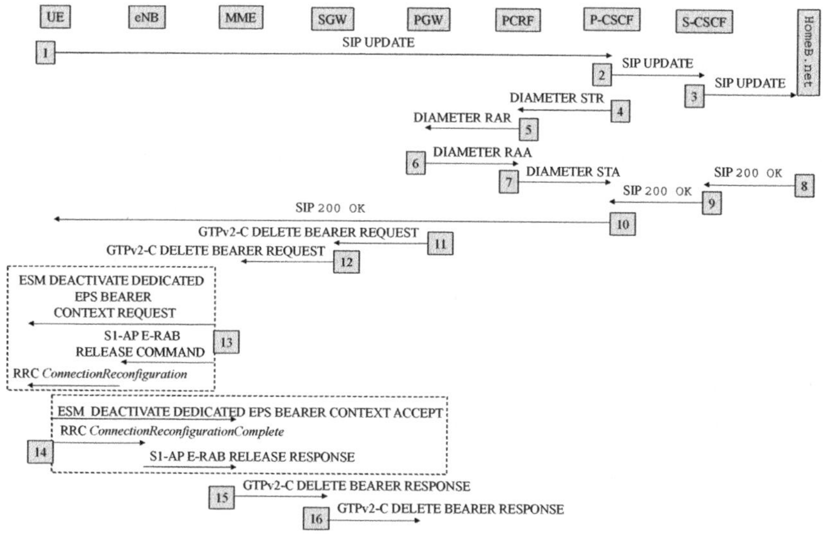

Figure 3.11. *Removing a video stream: initiated side*

The procedure of the removal of the video stream, at the received side, is described in Figure 3.12.

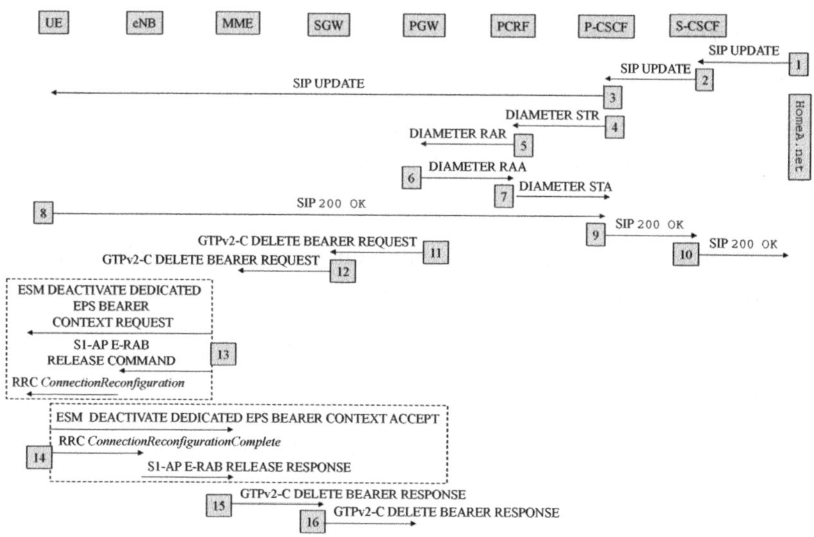

Figure 3.12. *Removing a video stream: received side*

1), 2), 3) Bob's UA entity (or Alice's) receives the UPDATE request, from the domain homeA.net (or homeB.net), indicating the removal of the video stream.

4) to 7) As described earlier, these messages are used to trigger the release of the dedicated for to the video stream.

8), 9), 10) Bob's UA entity (or Alice's) sends the 200 OK response to the UPDATE request to the domain homeA.net (or homeB.net), indicating the suppression of the video flow.

11) to 16) As described earlier, these messages are used to remove the dedicated bearer for the video stream.

3.9. Emergency call

If the mobile detects the emergency call, it performs a particular resource reservation with the EPS network to transmit the INVITE request.

The mobile will replace the number dialed by the user by the uniform resource name (URN) which defines the type of emergency call.

```
INVITE urn:service:sos SIP/2.0
...
```

The EPS network establishes a bearer (QCI = 5) whose allocation and retention priority (ARP) is higher than that established during the attachment of the mobile.

If the mobile does not detect the emergency call, the INVITE request is transmitted on the normal bearer (QCI = 5).

If the P-CSCF entity detects the emergency call, it can answer the mobile with an emergency call indication (Figure 3.13).

Upon receiving the INVITE message, the P-CSCF entity requests the PCRF entity for the ECGI identifier of the cell where the mobile is located.

The P-CSCF entity inserts that identifier in the INVITE request that it transfers to the Emergency CSCF entity (E-CSCF), in the following two cases (Figure 3.13):

– the mobile is not registered, because it has no UICC card or because it is on a visited network that has no roaming agreement with the home network;

– the mobile is already registered with the S-CSCF entity and a specific registration for the emergency call is not required.

If a specific registration for the emergency call is required, the P-CSCF entity informs the mobile (Figure 3.13).

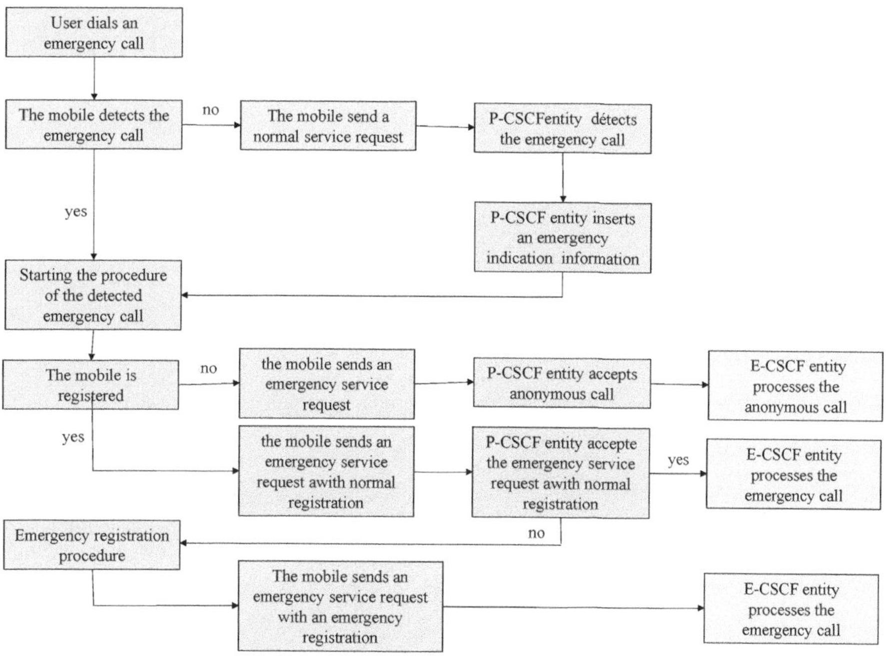

Figure 3.13. *Conditions for the transmission of the emergency call*

The REGISTER request shall insert in the header `Contact` the parameter `sos` to indicate that it is a registration for an emergency call.

```
REGISTER sip:ims.mnc01.mcc208.3gppnetwork.org SIP/2.0
...
Contact: <sip:192.0.2.101>; expires=600000; sos;
...
```

4

Radio Interface Procedures

4.1. Radio interface

At the LTE-Uu radio interface, between the user equipment (UE) and the evolved node base station (eNB), traffic data, corresponding to an IP Internet protocol (IP) packet and signaling data, corresponding to an radio resource control (RRC) message, are encapsulated by the data link layer broken down into three sub-layers (Figure 4.1):

– packet data convergence protocol (PDCP);

– radio link control (RLC) protocol;

– medium access control (MAC) protocol.

Three types of channels are defined (Figure 4.1):

– the logical channel defines the data structure at the interface between the RLC and MAC sub-layers;

– the transport channel defines the data structure at the interface between the MAC sub-layer and the physical layer;

– the physical channel defines the data structure between the two parts making up the physical layer, firstly, the coding and, secondly, the modulation and the multiplexing.

The RRC messages can carry non-access stratum (NAS) messages exchanged between the mobile and the mobility management entity (MME).

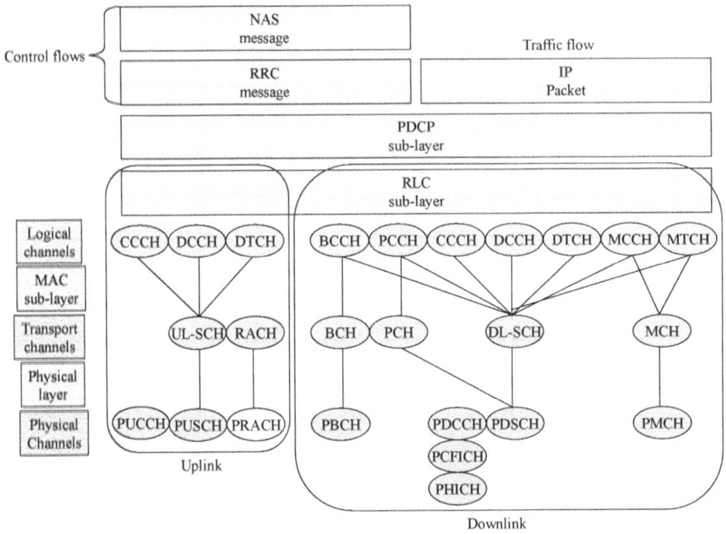

Figure 4.1. *Radio interface structure. For a color version of the figure, see www.iste.co.uk/perez/volte.zip*

4.1.1. *Data link sub-layer*

4.1.1.1. *PDCP protocol*

The PDCP protocol is used for RRC signaling messages, relating to dedicated control data and for the traffic IP packets.

The PDCP protocol performs the following functions:

– compression of the data traffic headers using the robust header compression (ROHC) mechanism;

– security of data traffic (confidentiality) and RRC signaling (integrity and confidentiality);

– the delivery in sequence of the RRC messages and the IP packets;

– the recovery of PDCP frames lost during the handover.

Several PDCP instances can be activated simultaneously:

– two instances for signaling radio bearers (SRB1 and SRB2) relating to RRC signaling data:

- the SRB1 bearer is used for the transmission of an RRC message that can carry a NAS message,

- the SRB2 bearer is used for the transmission of a NAS message only;

– one instance for each data radio bearer (DRB) relating to data traffic.

Header compression is based on the ROHC mechanism for which multiple algorithms have been defined by the request for comments (RFC) specifications of the Internet engineering task force (IETF) standards body (Table 4.1).

Profile identifier	*Compressed headers*	*References*
0×0000	uncompressed	RFC 4995
0×0001	RTP/UDP/IP	RFC 3095, RFC 4815
0×0002	UDP/IP	RFC 3095, RFC 4815
0×0003	ESP/IP	RFC 3095, RFC 4815
0×0004	IP	RFC 3843, RFC 4815
0×0006	TCP/IP	RFC 4996
0×0101	RTP/UDP/IP	RFC 5225
0×0102	UDP/IP	RFC 5225
0×0103	ESP/IP	RFC 5225
0×0104	IP	RFC 5225

Table 4.1. *ROHC specifications*

Header compression is based on the observation that in a session, a number of fields are invariant, such as IP addresses or port numbers.

Header compression is particularly effective when the size of the IP packet payload is relatively low, which is the case for voice (Figure 4.2).

The decompressor uses the PDCP feedback control message to inform the compressor that decompression is successful or that synchronization between compression and decompression has been lost.

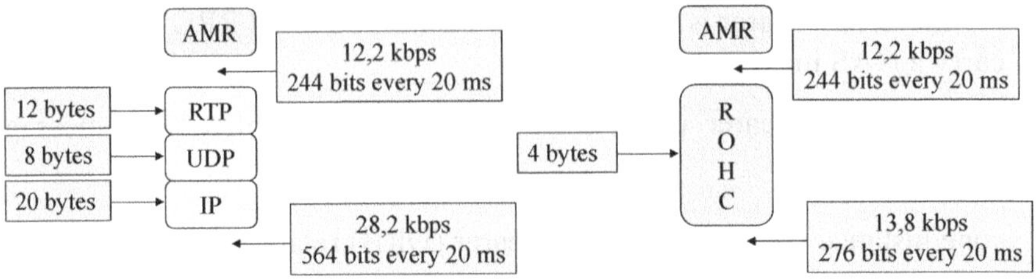

Figure 4.2. *Header compression*

4.1.1.2. *RLC protocol*

The RLC protocol provides control of the radio link between the mobile and the eNB entity.

The mobile can simultaneously activate multiple RLC instances, with each instance corresponding to a PDCP instance.

The RLC protocol operates in three modes:

– acknowledged mode (AM): session information protocol (SIP) uses this mode;

– unacknowledged mode (UM): real-time transport protocol (RTP) uses this mode.

– transparent mode (TM) for which no header is added.

The RLC protocol performs the following operations:

– retransmission in the case of error via the automatic repeat request (ARQ) mechanism, for the acknowledged mode only;

– concatenation, segmentation and reassembly of PDCP frames both in the acknowledged and unacknowledged mode;

– possible re-segmentation of PDCP frames, in the acknowledged mode, during retransmission of the RLC frame;

– re-sequencing of received data, both in the acknowledged and unacknowledged mode;

– detection of duplicate data both in the acknowledged and unacknowledged mode.

4.1.1.3. *MAC protocol*

The MAC protocol provides the following functions:

– multiplexing of RLC frames from multiple instances in a transport block;

– resource allocation via a scheduling mechanism;

– management of retransmission in case of error via the hybrid automatic repeat request (HARQ) mechanism;

– management of the random access procedure.

4.1.2. *Logical channels*

The broadcast control channel (BCCH) is a unidirectional common control channel, used only in the downlink for RRC broadcasting of master information block (MIB) and system information block (SIB) messages.

The paging control channel (PCCH) is a unidirectional common control channel, used only in the downlink to transport RRC messages for paging.

The common control channel (CCCH) is a bidirectional common control channel, used to transmit the first RRC signaling messages when the mobile connects to the eNB entity.

The dedicated control channel (DCCH) is a bidirectional dedicated control channel, used to transmit RRC messages when the mobile is connected to the eNB entity.

The dedicated traffic channel (DTCH) is a dedicated bidirectional channel, used to transmit unicast IP packets.

The multicast control channel (MCCH) is a unidirectional channel used for transmitting RRC messages for control information associated with IP packets transmitted in broadcast mode.

The multicast traffic channel (MTCH) is a unidirectional channel used to transmit IP packets in broadcast mode to the mobile.

4.1.3. *Transport channels*

4.1.3.1. *Downlink*

The broadcast channel (BCH) supports the BCCH logical channel including the RRC message relating to MIB system information.

The paging channel (PCH) supports the PCCH logical channel.

The downlink shared channel (DL-SCH) multiplexes the CCCH, DCCH, DTCH and BCCH logical channels. The BCCH logical channel includes the RRC messages relating to SIB system information.

The MCCH and MTCH logical channels are mapped to the DL-SCH transport channel if the number of mobiles concerned by transmitted IP packets in broadcast mode is low.

The multicast channel (MCH) multiplexes the MCCH and MTCH logical channels if the number of mobiles concerned by IP packets transmitted in broadcast mode is significant.

4.1.3.2. *Uplink*

The random access channel (RACH) does not transport logical channels. It is used by the mobile for random access to the eNB entity. The RACH transport channel only carries a preamble to initialize the connection with the eNB entity.

The uplink shared channel (UL-SCH) multiplexes the DCCH, CCCH and DTCH logical channels.

4.1.4. *Physical layer*

4.1.4.1. *Transmission chain*

The transmission chain consists of two sub-sets:

– for each direction of transmission, the first sub-set comprises the error detection code, the error correction code and the bit rate matching;

– for the downlink direction, the second sub-set includes modulation, mapping on spatial layers, precoding, resource element mapping and inverse fast Fourier transform (IFFT) to generate the orthogonal frequency division multiple access (OFDMA) signal (Figure 4.3).

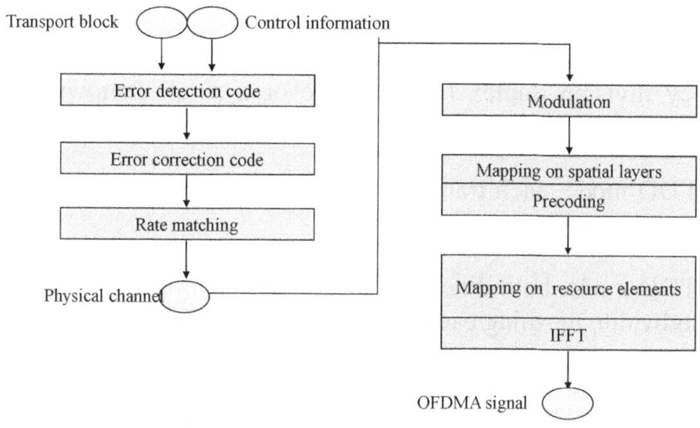

Figure 4.3. *Transmission chain: downlink*

– for the uplink direction, the second sub-set includes modulation, mapping to resource elements and inverse Fourier transform IFFT. The generation of the single carrier frequency division multiple access (SC-FDMA) signal introduced a fast Fourier transform (FFT). The spatial layer mapping and precoding are implemented only for Release 10 (Figure 4.4).

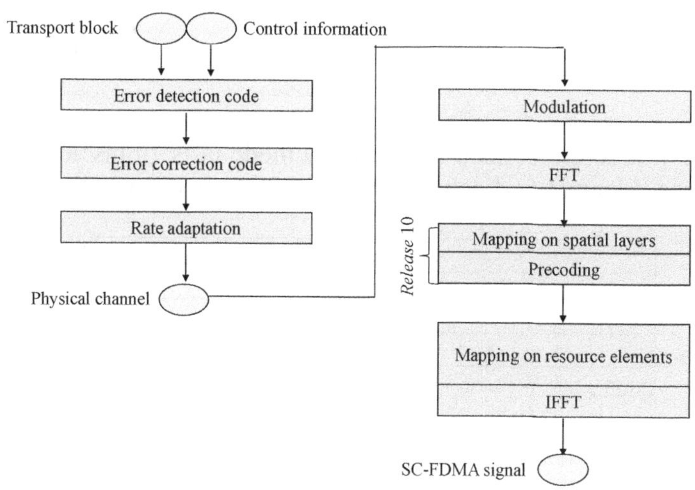

Figure 4.4. *Transmission chain: uplink*

4.1.4.2. *Frequency-division multiplexing*

Support for both transmission directions uses two bandwidths matched in the frequency division duplex (FDD) mode or a single bandwidth in the time-division duplex (TDD) mode.

For the FDD mode, each transmission direction operates simultaneously in the assigned bandwidth.

For the TDD mode, both directions of transmission operate alternately in the same bandwidth, meaning each direction is assigned a portion of time.

The bandwidth of the radio channel is flexible and can take several values: 1.4 MHz, 3 MHz, 5 MHz, 10 MHz, 15 MHz and 20 MHz.

Carrier aggregation (CA) involves combining the use of several component carriers (CC) or radio channels to increase the cell bit rate. The aggregation can be performed on five radio channels, bringing the maximum bandwidth to 100 MHz.

The radio channel is formed in the frequency domain of an orthogonal frequency-division multiplexing (OFDM) with sub-carrier spacing of 15 kHz or 7.5 kHz.

4.1.4.3. *Time-division multiplexing*

Two structures of time frames are defined according to the FDD or TDD mode.

The type-1 structure defined for the FDD mode lasts 10 ms and contains 10 sub-frames (Figure 4.5). Each sub-frame is made up of two time slots.

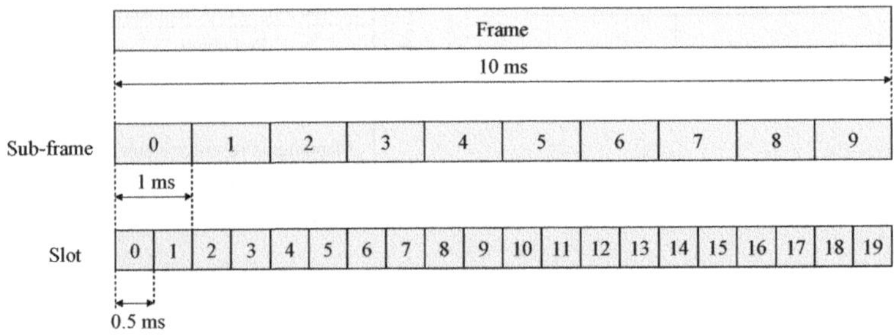

Figure 4.5. *Structure of the frame in FDD mode*

For the uplink, the signals transmitted by different mobiles must be temporally aligned, to the reception by the eNB entity.

The mobiles must, therefore, be synchronized temporally by the eNB entity which conveys the timing advance (TA) to them, to be applied for the uplink.

The type-2 structure defined for the TDD mode also lasts 10 ms and contains two half-frames of 5 ms each (Figure 4.6).

Each half-frame comprises five sub-frames and the second sub-frame can correspond to a special sub-frame containing three particular fields:

– a field for the downlink pilot time slot (DwPTS) which can contain data;

– a field for the uplink pilot time slot (UpPTS) which can contain data or a preamble;

– a gap period (GP), between the two preceding fields. This interval time facilitates the compensation of a time difference between different mobiles and avoids an overlap between the two transmission directions.

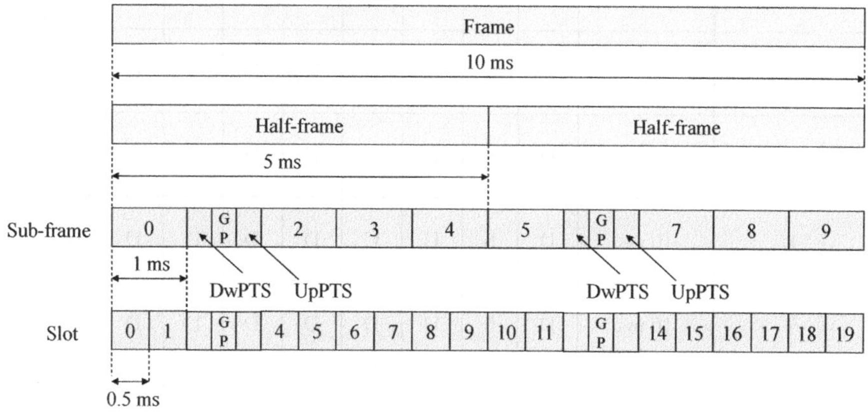

Figure 4.6. *Structure of the frame in TDD mode*

The sub-frames are attributed to the data for the uplink and downlink according to diverse configurations (Table 4.2):

– sub-frames 0 and 5 are always allocated to traffic in the downlink;

– sub-frame 1 is always allocated to the special sub-frame containing the three particular fields;

– sub-frame 2 is always allocated to traffic in the uplink;

– sub-frame 6 can be allocated to the special sub-frame containing three particular fields for a periodicity of 5 ms;

– sub-frames 3, 4, 7, 8, 9 are allocated to the downlink or uplink traffic according to the selected configuration.

Configuration	Periodicity	Number of the sub-frame									
		0	1	2	3	4	5	6	7	8	9
0	5 ms	D	S	U	U	U	D	S	U	U	U
1	5 ms	D	S	U	U	D	D	S	U	U	D
2	5 ms	D	S	U	D	D	D	S	U	D	D
3	10 ms	D	S	U	U	U	D	D	D	D	D
4	10 ms	D	S	U	U	D	D	D	D	D	D
5	10 ms	D	S	U	D	D	D	D	D	D	D
6	5 ms	D	S	U	U	U	D	S	U	U	D

D (*Downlink*) / U (*Uplink*) / S (*Special*)

Table 4.2. *TDD frame configuration*

Each time slot comprises of 3 or 6 or 7 OFDM symbols (Figure 4.7).

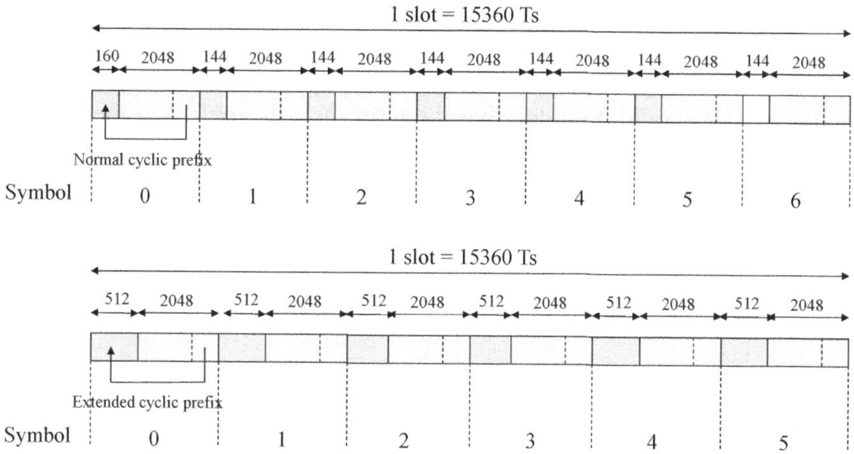

Figure 4.7. *Time slot structure*

An OFDM symbol corresponds to a number of bits which depends on the modulation used:

– 2 bits in the case of a quadrature phase-shift keying (QPSK) modulation whose constellation contains 4 different symbols;

– 4 bits in the case of a quadrature amplitude modulation (16-QAM) whose constellation contains 16 different symbols;

– 6 bits in the case of a 64-QAM modulation whose constellation contain 64 different symbols.

The introduction of a guard time between the symbols facilitates the removal of inter-symbol interference.

The guard time of a time slot is found at the beginning of the symbol. It contains the copy of the end of the symbol in order to avoid a very high dynamic for the amplifiers. This copy is called the cyclic prefix (CP).

The normal cyclic prefix is used in case the delay between the different reflected signals is low, which is the case in cells with a smaller diameter.

The extended cyclic prefix is used in case the delay between the different reflected signals is significant, which is the case in cells of a large diameter.

4.1.4.4. *Spatial multiplexing*

Four modes characterize the transmission system on the radio channel (Figure 4.8). It should be noted that the term input is applied to the input of the radio channel and the term output to the output of the same channel.

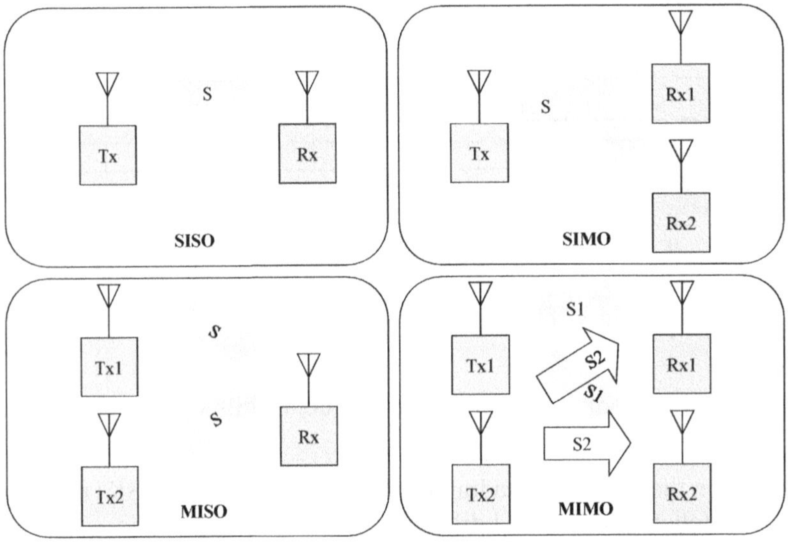

Figure 4.8. *Transmission modes*

The single input single output (SISO) mode is the basic signal propagation mode in which a transmitting and a receiving antenna are used.

The single input multiple output (SIMO) mode is characterized by the use of a single transmitting antenna and multiple receiving antennas. The SIMO mode is often referred to as receive diversity. The transmitted bit rate is identical to the SISO mode. On the other side, the selection of the received signal allows the receiver to protect against fading of the radio signal.

The multiple input single output (MISO) mode features multiple transmit antennas and a single reception antenna. The same signal is transmitted on the transmit antennas. The MISO mode is often referred to as transmit diversity. As for the SIMO mode, the MISO mode allows the receiver to protect against radio signal fading.

The MISO mode is also used for beamforming, directed towards the mobile, by controlling different phases of the emitted signals.

The multiple input multiple output (MIMO) uses multiple antennas for transmission and reception. It improves bit rate by allowing the transmission of several different signals on the same frequency and at the same time.

4.1.5. *Physical signals*

4.1.5.1. *Downlink*

The primary synchronization signal (PSS) ensures the frequency synchronization of the OFDMA signal and timing synchronization at the half-frame level.

The secondary synchronization signal (SSS) provides time synchronization at the frame level.

The cell-specific reference signal (RS) is a signal specific to the cell used to perform coherent demodulation of the received signal which is based on the calculation of the transfer function of the radio channel.

The MBMS single frequency network reference signal (MBSFN RS) is transmitted only in the physical multicast channel (PMCH) to perform coherent demodulation of the received signal.

The UE-specific RS physical signal is a specific signal to the mobile used to perform coherent demodulation of the received signal, to measure the power of the received signal and for beamforming.

The positioning reference signal (PRS) is used by the mobile to determine its location from the observed time difference of arrival (OTDOA) mechanism.

The channel state information reference signal (CSI RS) improves the measurement of the received signal and the interference level from that supplied from the cell-specific RS physical signal.

The power of the CSI RS physical signal is either transmitted to determine the level of the received signal, or suppressed to measure the level of interference.

4.1.5.2. *Uplink*

The demodulation reference signal (DM-RS) associated with the physical uplink shared channel (PUSCH) is used for estimation and synchronization of the PUSCH physical channel.

The DM-RS physical signal associated with the physical uplink control channel (PUCCH) is used for estimation and synchronization of the PUCCH physical channel.

The sound reference signal (SRS) allows the eNB entity to measure the quality of the signal for the uplink, in a frequency band higher than that allocated to the mobile.

This measurement cannot be obtained by the DM-RS physical signal because the DM-RS is associated with the PUSCH or PUCCH physical channels.

The measurement performed by the eNB entity allows it to set the frequency location of the resource allocated to the mobile for the uplink direction, and the modulation and coding scheme.

4.1.6. *Physical channels*

4.1.6.1. *Downlink*

The physical broadcast channel (PBCH) transmits the BCH transport channel containing the system information corresponding to the MIB message.

The physical control format indicator channel (PCFICH) transmits the control format indicator (CFI) indicating the size of the physical downlink control channel (PDCCH).

The physical HARQ indicator channel (PHICH) transmits the HARQ indicator (HI) which indicates a positive (ACK) or negative (NACK) acknowledgment for the uplink data received, in the physical uplink shared channel (PUSCH).

The PDCCH physical channel transmits downlink control information (DCI):

– allocation of resources and the modulation and coding scheme, for the data in the PDSCH and PUSCH physical channels;

– transmission power of the PUCCH and PUSCH physical channels.

The physical downlink shared channel (PDSCH) transmits the DL-SCH and PCH transport channels.

The physical multicast channel (PMCH) transmits the MCH transport channel.

4.1.6.2. *Uplink*

The physical random access channel (PRACH) contains a preamble used by the mobile when it needs to perform a random access, which is the first stage of the connection of the mobile to the eNB entity.

The PUCCH physical channel uses three types of format to transport the uplink control information (UCI):

– formats 1, 1a and 1b transport the UCI information relating to the scheduling request to obtain resource on the PUSCH physical channel and to the positive (ACK) or negative (NACK) acknowledgment, corresponding to the HARQ mechanism, for the data received on the PDSCH physical channel;

– formats 2, 2a and 2b transport UCI information relating to the signal status reports for the signal received on the PDSCH physical channel and the positive (ACK) or negative (NACK) acknowledgments;

– format 3 transports the same information as format 1 by adapting it to the aggregation of radio channels introduced in Release 10.

The PUSCH physical channel transmits the UL-SCH transport channel and the UCI control information.

For Releases 8 and 9, the simultaneous transmission of PUSCH and PUCCH physical channels is not supported.

The transmission of UCI information in the PUSCH physical channel is carried out, on the one hand, together with the transfer of traffic data or RRC control, while for the transfer of aperiodic UCI information reports, on the other.

For Release 10, the simultaneous transmission of the PUSCH and PUCCH physical channels is supported.

The transmission of UCI information in the PUCCH physical channel is maintained when the data traffic or RRC control need to be transferred to the PUSCH physical channel.

4.2. Procedures

4.2.1. *Access control*

4.2.1.1. *Acquisition of the PRACH physical channel*

The procedure of access to the eNB entity is developed for accessing the physical random access channel (PRACH) that the mobile is going to use to carry out the random access.

The different physical channels and physical signals processed by the mobile for the acquisition of the PRACH physical channel can be found in Table 4.3.

Physical channels Physical signals	*Acquisition of the parameters*
PSS	Frequency synchronization Time synchronization Parameter $N_{\text{ID}}^{(2)}$
SSS	Time synchronization Length of the cyclic prefix Parameter $N_{\text{ID}}^{(1)}$
Cell-Specific RS	Coherent demodulation
PBCH	Bandwidth of the channel. for the downlink Parameter of the PHICH physical channel
PCFICH	Size of the PDCCH physical channel
PDCCH	Detection of the SI-RNTI identity
PDSCH	SIB1 and SIB2 system information

Table 4.3. *Acquisition of the PRACH physical channel*

The PSS physical signal facilitates the following functions:

– frequency synchronization;

– time synchronization at the level of the OFDM symbol, of the time slot, of the sub-frame (1-ms periodicity) and the half-frame (5-ms periodicity);

– determination of the value of the number $N_{\text{ID}}^{(2)}$.

The SSS physical signal facilitates the following functions:

– time synchronization at the frame level (periodicity 10 ms);

– determination of the FDD or TDD mode;

– determination of the Cyclic Prefix (CP) type, normal or extended;

– determination of the value of the number $N_{\text{ID}}^{(1)}$.

The number of the physical-layer cell identity (PCI) is $N_{\text{ID}}^{\text{cell}} = 3N_{\text{ID}}^{(1)} + N_{\text{ID}}^{(2)}$.

$N_{\text{ID}}^{(1)}$ represents the group number and can take the values between 0 and 167. It is determined by the SSS physical signal.

$N_{\text{ID}}^{(2)}$ represents the number in the group and can be a value between 0 and 2. It is determined by the PSS physical signal.

The value $N_{\text{ID}}^{\text{cell}}$ determines the mapping of the cell-specific (RS) physical signal on the resource elements (RE).

A resource element corresponds to a symbol in the time domain and one sub-carrier in the frequency domain.

After having withdrawn the resource elements allocated to the cell-specific RS physical signal, the mobile can analyze the PBCH physical channel which transmits the BCH transport channel containing the system information corresponding to the MIB message.

The MIB message provides the information which allows the mobile to analyze the PCFICH and the PDCCH physical channels subsequently:

– the bandwidth of the radio channel, for the downlink;

– the system frame number (SFN);

– the configuration of the PHICH physical channel.

The processing of the PSS and SSS physical signal, on the one hand, and of the PBCH physical channel, on the other hand, is independent of the bandwidth of the radio channel.

After having withdrawn the resource elements allocated to the cell-specific RS physical signal, the mobile can analyze the PCFICH physical channel which, for each sub-frame, defines the number of OFDM symbols allocated to the PDCCH physical channel.

After having withdrawn the resource elements allocated to the cell-specific RS physical signal, to the PCFICH and PHICH physical channels, the mobile can analyze the PDCCH physical channel.

The PDCCH physical channel transports the information which facilitates the recovery of the System Information Base 1 and 2 (SIB1 and SIB2) messages contained in the PDSCH physical channel from the detection of the system information radio network temporary identifier (SI-RNTI).

The SIB1 message provides the following information:

– the bandwidth of the radio channel, for the uplink;

– the configuration of the type-2 time frame;

– the scheduling of the other SIB system information.

The SIB2 message provides the information related to the configuration of radio resources allocated to the PRACH physical channel.

4.2.1.2. *Random access*

The random access procedure is initialized by the mobile and is required by the following cases:

– when establishing the connection to the eNB entity;

– when changing the cell during the session (handover);

– when updating timing advance (TA);

– when re-establishing the connection to the eNB entity.

The random access procedure is said to be with contention when the mobile chooses the used resource (PRACH channel, preamble) randomly. This occurs for the establishment or re-establishment phases of the connection.

The random access procedure is said to be without contention when the eNB entity is providing the resource to be used to the mobile. This happens for the handover or the updating of the timing advance TA.

4.2.1.2.1. Random access with contention

The random access procedure with contention, during the establishment or re-establishment of the connection to the eNB entity is described in Figure 4.9.

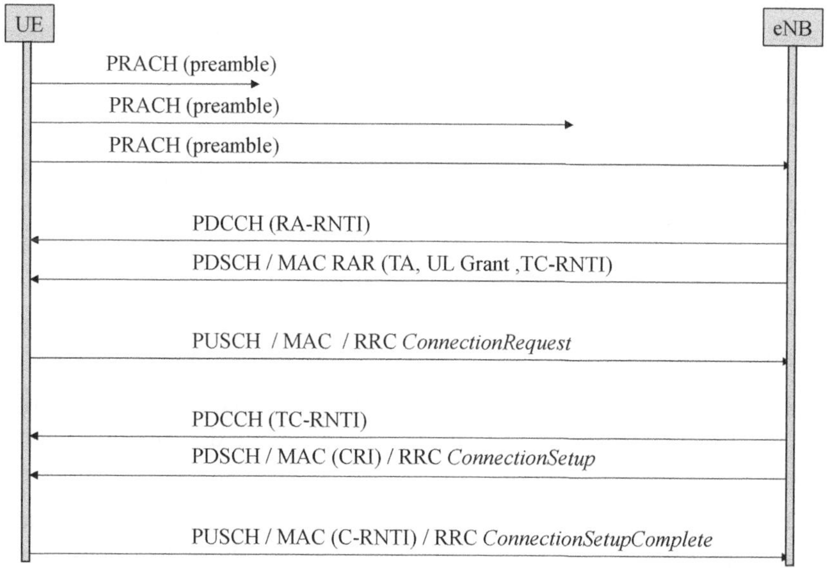

Figure 4.9. *Random access with contention*

The mobile transmits the preamble in the PRACH physical channel. This preamble is chosen randomly in a list communicated by the eNB entity in the SIB2 system information.

In case the eNB entity does not respond, the mobile retransmits the preamble while increasing the transmission power. The maximum number of retransmissions is shown by the SIB2 system information or the RRC *Connection Reconfiguration* message.

The risk of contention is linked to the fact that several mobiles can access the same PRACH physical channel and use the same preamble.

When the eNB entity receives the PRACH physical channel, it calculates the timing advance and transmits to the mobile:

– the DCI control information in the PDCCH physical channel, recovered from random access RNTI (RA-RNTI) identity. The mobile retrieves the description of its data in the PDSCH physical channel;

– the random access response (MAC RAR) frame containing the index of the preamble, the timing advance TA, the resource to use (UL Grant) for the transmission in the PUSCH physical channel and the temporary cell RNTI (TC-RNTI) identity. This identity is temporary as several mobiles may consider that this identity is allocated to them, thereby leading to contention.

The mobile initializes its timing advance and responds with the RRC *Connection Request* message containing:

– the shortened temporary mobile subscriber identity (S-TMSI), if the mobile is already attached;

– a random number in the contrary case.

When the eNB entity receives the RRC *Connection Request* message, it transmits to the mobile:

– the DCI control information in the PDCCH physical channel, recovered from the TC-RNTI identity. The mobile retrieves the description of its data in the PDSCH physical channel;

– the header MAC RAR containing the UE CRI (contention resolution identity) control element. This control element reproduces the identity in the RRC *Connection Request* message, thereby resolving the contention issue;

– the RRC *ConnectionSetup* message.

The TC-RNTI identity becomes the C-RNTI, the definitive identity allocated to mobile.

The mobile indicates its C-RNTI in the control element of the MAC frame and confirms the connection through the RRC *ConnectionSetupComplete* message.

4.2.1.2.2. Random access without contention

The random access procedure without contention, when changing the cell during the session, is described in Figure 4.10.

During the procedure of handover based on the X2 interface, the target eNB entity provides the characteristics of the radio interface to the source eNB entity in a message X2-AP HANDOVER REQUEST ACK.

This message contains the information element *handover command* which specifies the preamble that the mobile must use during the random access procedure to the target eNB entity.

The source eNB entity forwards the information element *handover command* in the RRC *Connection Reconfiguration* message which triggers the handover for the mobile.

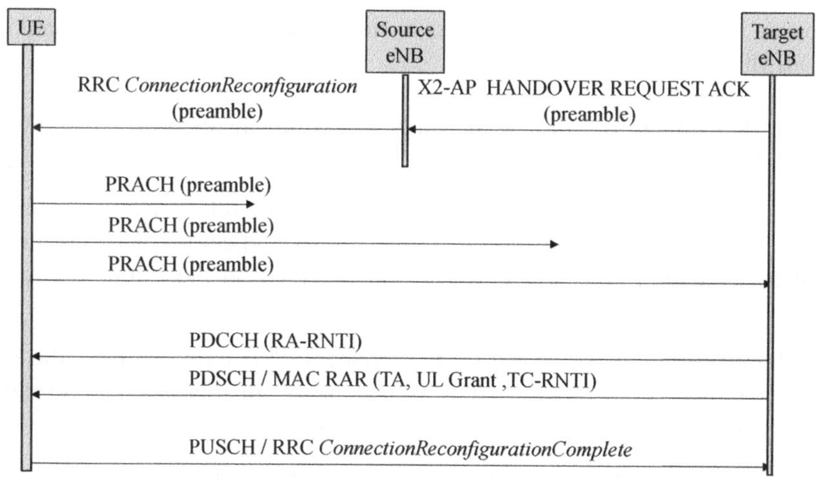

Figure 4.10. *Random access without contention in case of changing the cell during the session*

The random access procedure comprises as before the preamble transmission by the mobile and the MAC RAR frame by the eNB entity.

The mobile connection is finalized when the mobile transmits the RRC *Connection Reconfiguration Complete* message.

4.2.2. Data transfer

4.2.2.1. Scheduling

Data scheduling is the operation carried out by the eNB entity which consists of providing resource blocks (RB) to the mobiles and in controlling transmission power, for the downlink and the uplink direction.

In the time domain, the allocation of the resources corresponds to a sub-frame of a TTI (Transmission Time Interval) duration of a millisecond, which represents two resource blocks (RB).

In the frequency domain, the eNB entity can give the mobile several resource blocks, each block corresponds to a frequency band of 180 kHz, formed from an orthogonal frequency division multiplexing (OFDM) of 12 sub-carriers spaced 15 kHz or 24 sub-carriers spaced 7.5 kHz.

In the space domain, the mobile can receive and transmit different resource blocks, simultaneously and in the same frequency band, thanks to the MIMO mechanism.

For the downlink, the scheduling algorithm takes into account the following information:

– the information recovered by the mobiles, in relation to the channel quality indicator (CQI), to the precoding matrix indicator (PMI) and to the number of spatial layers from the signal received on the PDSCH physical channel;

– the information recovered by the adjacent eNB entities under the inter cell interference coordination (ICIC) mechanism, in relation to the relative narrowband tx power (RNTP) power transmission;

– the information transmitted by the MME entity in relation to the category of the mobile and to the QoS class identifier (QCI) level;

– the local information related to the state of the memory, to the need of retransmission, to the state of available resources and to the measure intervals constructed for the mobile.

For the uplink, the scheduling algorithm takes into account the following information:

– the information recovered by the mobiles, in relation to the power headroom report (PHR) and the buffer status report (BSR);

– the information recovered by the adjacent eNB entities under the ICIC mechanism, in relation to the interference overload indication (IOI) and to high interference indication (HII);

– the information recovered by the MME entity in relation to the category of the mobile and to the QCI level;

– the local information related to the level of quality measured on the SRS physical signal, to the need for retransmission, to the state of available resources and to the measure intervals constructed for the mobile.

The results of the scheduling make up the DCI control information communicated in the PDCCH physical channel.

The DCI control information in formats 0 and 4 facilitates the scheduling of the data in the PUSCH physical channel.

The DCI control information in formats 1, 1A, 1B, 1C, 1D and 2, 2B, 2C facilitates the scheduling of data in the PDSCH physical channel.

The DCI control information in format 3 facilitates the transmit power control (TPC) of the PUSCH and PUCCH physical channels.

The allocation of the resources is shown by the radio network temporary identifier (RNTI) specific to a mobile (C-RNTI, SPS C-RNTI and TPC-RNTI) or to a set of mobiles (P-RNTI, RA-RNTI, TC-RNTI and SI-RNTI).

In the FDD mode, the resource allocation for the uplink, signaled in the PDCCH physical channel of the sub-frame n, is applicable to the sub-frame n+4 ms.

In the TDD mode, the shift between signaling in the PDCCH physical channel and transmission in the PUSCH physical channel depends on the configuration of the time frame and the number of the sub-frame of the PDCCH physical channel (Table 4.4).

Configuration of the time frame	Number of the sub-frame of the PDCCH physical channel									
	0	1	2	3	4	5	6	7	8	9
1	–	6	–	–	4	-	6	–	–	4
2	–	–	–	4	–	–	–	–	4	–
3	4	–	–	–	–	–	–	–	4	4
4	–	–	–	–	–	–	–	–	4	4
5	–	–	–	–	–	–	–	–	4	
6	7	7	–	–	–	7	7	–	–	5

Table 4.4. *Shift between signaling in the PDCCH channel and transmission in the PUSCH channel*

For configuration 0 of the time frame, the shift between signaling in the PDCCH physical channel and transmission in the PUSCH channel depends on the value of the two bits Uplink Index of DCI control information in formats 0 and 4:

– for the most significant bit at ONE and the least significant bit at ZERO, the shift is provided in Table 4.5;

– for the most significant bit at ZERO and the least significant bit at ONE, the shift is fixed at 7 ms;

– for the most significant bit at ONE and the least significant bit at ONE, the mobile can transmit in the two sub-frames previously defined.

Configuration of the time frame	Number of the sub-frame of the PDCCH physical channel									
	0	1	2	3	4	5	6	7	8	9
0	4	6	–	–	–	4	6	–	–	–

Table 4.5. *Shift for configuration 0 of the time frame*

4.2.2.2. *DRX function*

The discontinuous reception (DRX) function determines the moments when the mobile must analyze the PDCCH physical channel, which allows it to avoid processing this channel every millisecond and in this way to preserve the consumption of its battery (Figure 4.11).

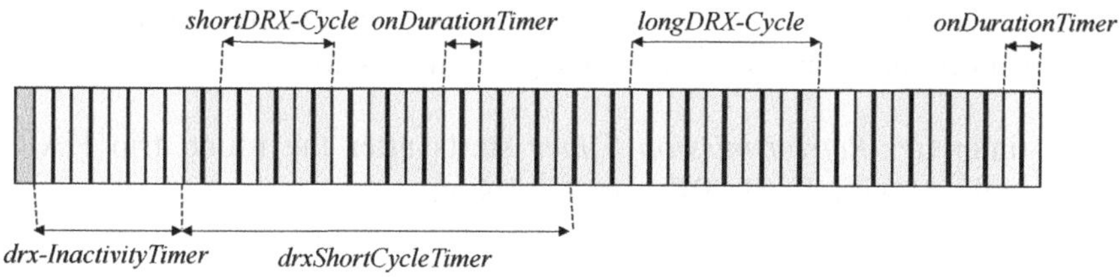

Figure 4.11. *DRX function For a color version of the figure, see www.iste.co.uk/perez/volte.zip*

An inactivity timer *drx-Inactivity Timer* of the DRX function is triggered when the mobile receives data on the PDCCH physical channel.

This timer is reinitialized each time that the mobile receives data on the PDCCH physical channel.

When the timer *drx-Inactivity Timer* expires, the mobile starts an optional period for the short cycle corresponding to the timer *drxShort Cycle Timer*.

During the short cycle, the mobile analyzes the PDCCH physical channel in a duration corresponding to the timer *onDurationTimer*.

During the short cycle, the triggering of the active period is provided by the following formula:

[(SFN * 10) + (sub-frame number)] modulo (*shortDRX-Cycle*) = (*drxStartOffset*) modulo (*shortDRX-Cycle*)

When the timer *drxShort Cycle Timer* expires, the mobile starts the long cycle period for which the triggering of the active period is provided by the following formula:

[(SFN * 10) + (sub-frame number)] modulo (*longDRX-Cycle*) = *drxStartOffset*

The parameter configuration of the DRX function is indicated in the RRC messages of establishment or re-establishment of the connection.

The eNB entity indicates the activation of the DRX function while using the DRX control element in the MAC frame.

The timer *drx-Retransmission Timer* defines an inactive period of the DRX function during an expected HARQ retransmission.

The timer *drx-Retransmission Timer* starts after the round trip HARQ timer associated with the HARQ retransmission.

4.2.2.3. *SPS function*

The semi-persistent scheduling (SPS) function is applied to the applications that have a periodic character, for example the voice which produces a block every 20 milliseconds.

The SPS function allows the eNB entity to avoid announcing the DCI control information in the PDCCH physical channel.

The SPS function is configured by the eNB entity in RRC messages of establishment or re-establishment of the connection or establishment of Data Radio Bearer (DRB) containing the following parameter:

– SPS C-RNTI identifier;

– allocation periodicity, for example a transmission every 20 sub-frames for the voice;

– number of the HARQ process of retransmission in case of errors, for the downlink.

When the SPS function has been configured, the eNB entity uses the DCI control information transmitted in the PDCCH physical channel to activate or release it.

When the SPS function has been configured, the mobile must, nevertheless, continue the analysis of the PDCCH physical channel, for the defined sub-frames by the DRX function to detect the DCI control information in relation to the allocation release.

When the SPS function has been activated for the downlink, the sub-frame allocation of the PDSCH physical channel corresponds to the following formula:

(10 * SFN + sub-frame) = [(10 * *SFNstart time* + *subframestart time*) + N * *semiPersistSchedIntervalDL*] modulo 10240

The *SFN start time* and *sub frame start time* parameters correspond to the values of the frame and sub-frame number when the SPS function has been activated.

The *semiPersistSchedIntervalDL* parameter corresponds to the allocation periodicity of resources for the downlink.

When the SPS function has been activated for the uplink, the sub-frame allocation of the PUSCH physical channel corresponds to the following formula:

(10 * SFN + sub-frame) = [(10 * *SFNstart time* + *subframestart time*) + N * *semiPersistSchedIntervalUL* + *Subframe_Offset* * (N modulo 2)] modulo 10240

The *semiPersistSchedIntervalUL* parameter corresponds to the allocation periodicity of resources for the uplink.

The *Subframe_Offset* parameter is optional and its value is equal to ZERO for the FDD mode and is indicated in Table 4.6 for the TDD mode.

Configuration of the sub-frame	Number of the subf-rame during the SPS activation	Sub-frame_Offset
0	sans objet	0
1	2 and 7	1
	3 and 8	-1
2	2	5
	7	-5
3	2 and 3	1
	4	-2
4	2	1
	3	-1
5	sans objet	0
6	sans objet	0

Table 4.6. *Value of optional parameter Subframe_Offset*

4.2.2.4. *HARQ function*

In case of error, the retransmission implements two mechanisms:

– the automatic repeat request (ARQ) mechanism elaborated by the RLC layer;

– the hybrid ARQ (HARQ) mechanism established at the physical layer level, under the MAC layer control.

The ARQ mechanism takes over from the HARQ mechanism if the retransmissions at the physical layer level have failed.

The ARQ is applied only to the flow of traffic and signaling using the acknowledged mode (AM) of the RLC protocol.

On the other hand, taking its speed into consideration, the HARQ mechanism is applied to the flow using the AM mode and unacknowledged mode (UM) of the RLC protocol.

A different redundancy version facilitates the retransmission of selected data:

– for the chase combining mechanism, the retransmission contains the same sequence. The improvement of the error corrector code is due to the increase of the signal to noise ratio;

– for the incremental redundancy mechanism, the retransmission contains some different sequences. The improvement of the error corrector code is due to the increase of the number of redundancy bits.

The first transmission corresponding to the redundancy version (RV0) contains the initial sequence to transmit and redundancy bits generated by the error corrector code, whose number is determined by the coding rate.

The HARQ mechanism functions with a window of one data unit. When a transport block is transmitted, the transmitter must wait for the reception of the acknowledgement before sending the following transport block.

This disposition penalizes the rate of the mobile and this restriction is removed thanks to the implementation of several HARQ parallel processes.

The combination of HARQ parallel processes and of the retransmission mechanism provokes a de-sequencing of the blocks received by the destination. The RLC layer assures the re-sequencing of different blocks received.

For the downlink, the HARQ mechanism is adaptive because the retransmission can modify the transmission parameters.

For the downlink, the HARQ mechanism is asynchronous because a HARQ process transmission is not constrained by an imposed period.

For the uplink, the HARQ mechanism is adaptive or non-adaptive. It is non-adaptive if the retransmission must use the same initial transmission parameters.

For the uplink, the HARQ mechanism is synchronous because the HARQ process transmission must be done according to a defined timing.

4.2.2.4.1. Data transfer for uplink

The number of HARQ entities for transmission mode 2 (mode with MIMO) is double that of transmission mode 1 (mode without MIMO), because one HARQ entity is allocated to each transport block:

– one transport block is used for transmission mode 1;

– two transport blocks are used for transmission mode 2.

In case of the carrier aggregation (CA), a HARQ entity is developed for each radio channel.

For the FDD mode, by the HARQ entity, the HARQ process number has a fixed value:

– value equal to 8 for transmission mode 1;

– value equal to 16 for transmission mode 2.

Figure 4.12 described the HARQ mechanism for the transmission mode 1 in the FDD mode.

The transmission moments for each HARQ process is synchronous, with an 8 ms periodicity.

The PHICH physical channel transports the HARQ indicator (ACK or NACK bits) for each transport block, 4 milliseconds after the transmission of the data in the PUSCH physical channel.

Under a HARQ process, if a transport block is acknowledged, a new transport block can be transmitted 8 milliseconds later. If a transport block is not acknowledged, the RV1 redundancy version is transmitted 8 milliseconds later.

The HARQ processes used depend on the transmission moment of transport blocks. The 8 HARQ processes are systematically used if the eNB entity allocates every millisecond of resources to the mobile.

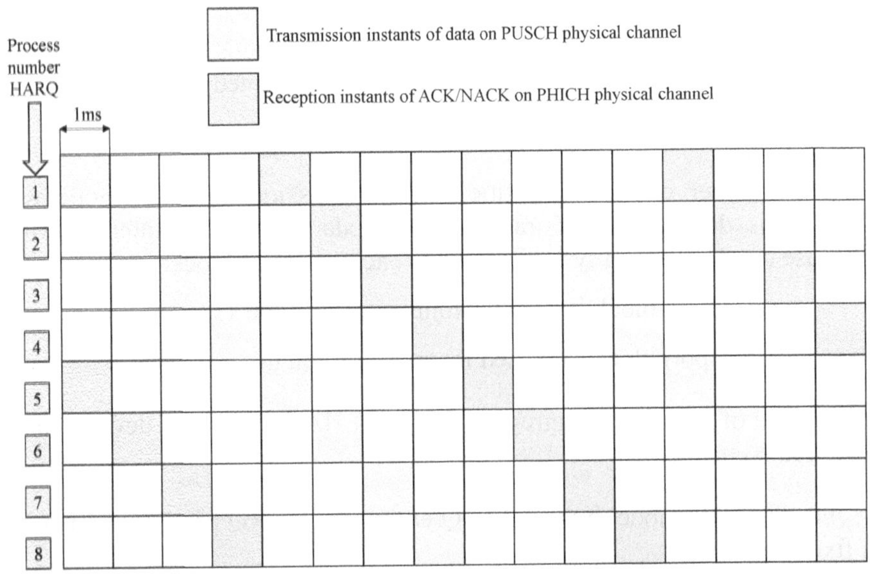

Figure 4.12. *HARQ function in the FDD mode Data transfer for uplink. For a color version of the figure, see www.iste.co.uk/perez/volte.zip*

For the TDD mode, the HARQ process number, under the HARQ entity, depends on the type of configuration of the time frame and transmission mode (Table 4.7).

Configuration of the time frame	0	1	2	3	4	5	6
Number of sub-frames / uplink	6	4	2	3	2	1	5
Transmission mode 1	7	4	2	3	2	1	6
Transmission mode 2	14	8	4	3	4	2	12

Table 4.7. *HARQ process number in the TDD mode Data transfer for uplink*

The shift which is given in milliseconds, between the transmission of a block in the PUSCH physical channel and the reception of the HI indicator in the PHICH physical channel, is shown in Table 4.8.

Configuration of the time frame	Number of the sub frame having received HI									
	0	1	2	3	4	5	6	7	8	9
0	7/6	4	–	–	–	7/6	4	–	–	–
1	–	4	–	–	6	–	4	–	–	6
2	–	–	–	6	–	–	–	–	6	–
3	6	–	–	–	–	–	–	–	6	6
4	–	–	–	–	–	–	–	–	6	6
5	–	–	–	–	–	–	–	–	6	–
6	6	4	–	–	–	7	4	–	–	6

Table 4.8. *Shift between the PUSCH and PHICH physical channels*

Figure 4.13 describes the HARQ mechanism for configuration 1 of the time frame in the TDD mode.

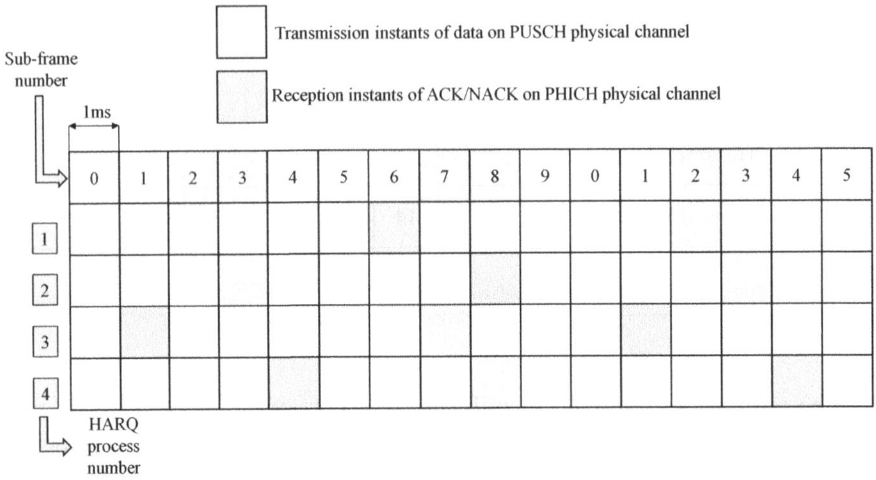

Figure 4.13. *HARQ function in the TDD mode, for configuration 1 Data transfer for uplink. For a color version of the figure, see www.iste.co.uk/perez/volte.zip*

The transport block transmitted in sub-frame 2 or 7 (sub-frame 3 or 8 respectively) receives the HI indicator with a shift of 4 milliseconds (6 milliseconds respectively).

Under a HARQ process, if a transport block is acknowledged, a new transport block can be transmitted 10 milliseconds later. If a transport block is not acknowledged, the RV1 redundancy version is transmitted 10 milliseconds later.

4.2.2.4.2. Data transfer for downlink

Contrary to the uplink, a single HARQ entity is developed independently of the transmission mode.

In case of the aggregation of radio channels, a HARQ entity is developed for each radio channel.

As the process number is shown in the DCI control information, the number of open HARQ processes is not fixed and depends on transmission needs.

For the FDD mode, the number of open HARQ processes is less than or equal to 8. The eight HARQ processes are only open if the eNB entity allocates resources to the mobile every millisecond.

The PUCCH or PUSCH physical channel transports the ACK or NACK indicator with a shift of 4 milliseconds after the transmission of the transport block in the PDSCH physical channel.

Under a HARQ process, if a transport block is acknowledged, a new transport block can be transmitted. If a transport block is not acknowledged, the RV1 redundancy version is transmitted. In the two cases, the moment used by the HARQ process is not imposed by timing as is the case for the uplink.

For the TDD mode, the HARQ process number, by the HARQ entity depends on the configuration of the time frame (Table 4.9).

Configuration of the time frame	0	1	2	3	4	5	6
Number of sub-frames downlink	4	6	8	7	8	9	5
HARQ process number	≤ 4	≤ 7	≤ 10	≤ 9	≤ 12	≤ 15	≤ 6

Table 4.9. *HARQ process number in the TDD mode data transfer for the downlink*

Table 4.10 shows the number of ACK/NACK bits to be transmitted, through the HARQ entity in the PUCCH physical channel, depending on the configuration of the time frame, for transmission mode 1. The indicated values are doubled in the case of transmission mode 2.

Configuration of the time frame	0	1	2	3	4	5	6
Number of sub-frames for the PUCCH physical channel	6	4	2	3	2	1	5
Number of ACK / NACK bits	4	6	8	7	8	9	5
Number of bits through sub-frame	1 ou 0	1 ou 2	4	2 ou 3	4	9	1

Table 4.10. *Number of ACK / NACK bits to be transmitted in the PUCCH physical channel for the TDD mode and the transmission mode 1*

In certain cases, the capacity of the PUCCH physical channel is insufficient for recovering all the ACK / NACK information:

– format 1a has a capacity of one bit;

– format 1b has a capacity of 2 or 4 bits.

Two methods are defined to reduce the number of bits:

– coupling carried out from a logical AND of various ACK / NACK information:

- the NACK information of a transport block involves the retransmission of all the transport blocks which are linked to it,

- in case an ACK / NACK bit is generated by a sub-frame, the coupling generates a bit transmitted in format 1a,

- in case two ACK / NACK bits are generated by the sub-frame, the coupling generates two bits transmitted in format 1b (Figure 4.14);

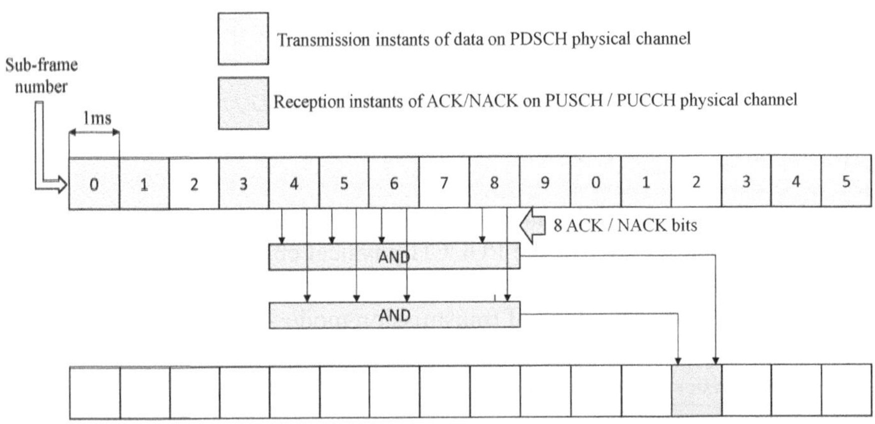

Figure 4.14. *Coupling ACK / NACK information Configuration 2 of the time frame Data transfer for downlink. For a color version of the figure, see www.iste.co.uk/perez/volte.zip*

– multiplexing of ACK / NACK information in the PUCCH physical channel:

- format 1b is used for transporting 4 bits of information, each bit corresponding to ACK / NACK information generated by a sub-frame,

- in case two bits are generated by the sub-frame, the two ACK / NACK corresponding bits are linked by a logical AND (Figure 4.15),

- the NACK information of a transport block of a sub-frame involves the retransmission of the other transport block of the same sub-frame.

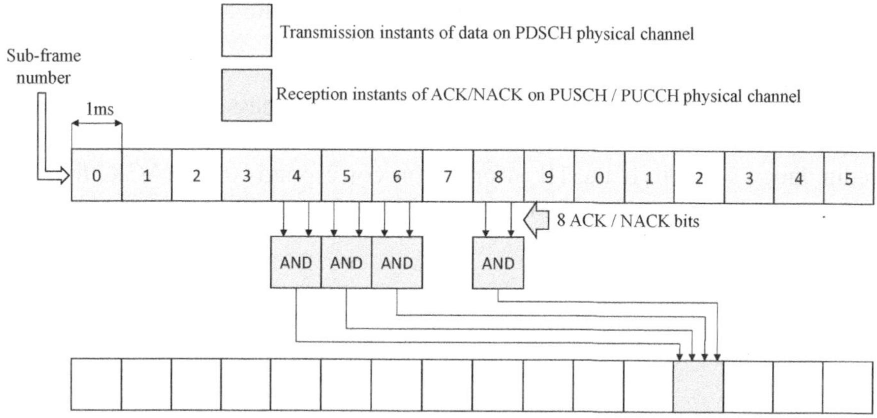

Figure 4.15. *Multiplexing ACK / NACK information Configuration 2 of the time frame data transfer for uplink. For a color version of the figure, see www.iste.co.uk/perez/volte.zip*

4.2.2.5. *TTI bundling function*

Real-time applications like the voice are sensitive to the packet jitter. The RTP protocol facilitates the correction of the jitter introduced in the network, with a maximum value of 40 ms and packets having a jitter greater than this value are suppressed.

The retransmission mechanism of the HARQ function results in an increase of the packet jitter for each retransmission (for example 8 ms in the FDD mode). The HARQ mechanism may be counter-productive in this case.

The TTI bundling function consists of transmitting, in four consecutive sub-frames, four redundancy versions, without waiting for the return of HI information, in order to reduce the value of the jitter.

The same modulation and coding scheme and the same frequency band are used for transmission of the four redundancy versions.

The TTI bundling function is limited to the transmission for the uplink.

The TTI bundling function is not supported for the aggregation of radio channels.

In the FDD mode, the number of HARQ processes is reduced to 4 (Figure 4.16). The HI information is transmitted 4 ms after the last redundancy version. If the HI information corresponds to a NACK, the four redundancy versions are retransmitted with a 16 ms delay.

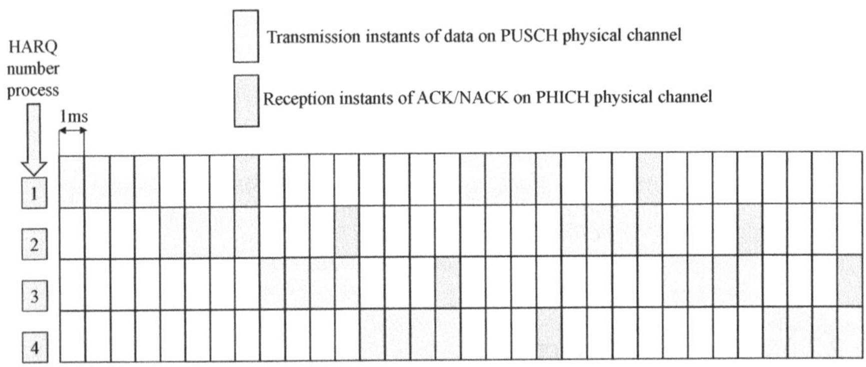

Figure 4.16. *TTI bundling function in the FDD mode*
For a color version of the figure, see
www.iste.co.uk/perez/volte.zip

In the TDD mode, the TTI bundling function is only applicable in configurations 0, 1 and 6 of the time frame.

The HARQ processes number is limited to 3 for the configurations 0 and 6 and to 2 for the configuration 1 (Figure 4.17).

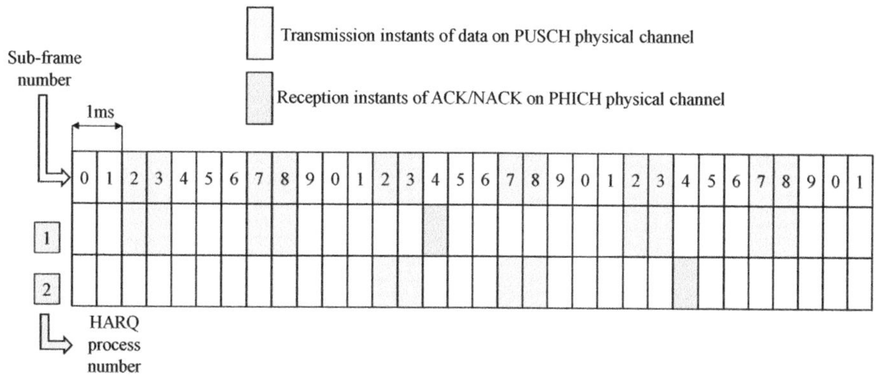

Figure 4.17. *TTI function bundling in the TDD mode for configuration 1. For a color version of the figure, see www.iste.co.uk/perez/volte.zip*

5

Service Profiles

5.1. Subscription data

5.1.1. *Subscription to the EPS network*

When subscribing to the evolved packet system (EPS), the user is assigned a private international mobile subscriber identity (IMSI) which is associated with a profile on the telephone service (Figure 5.1).

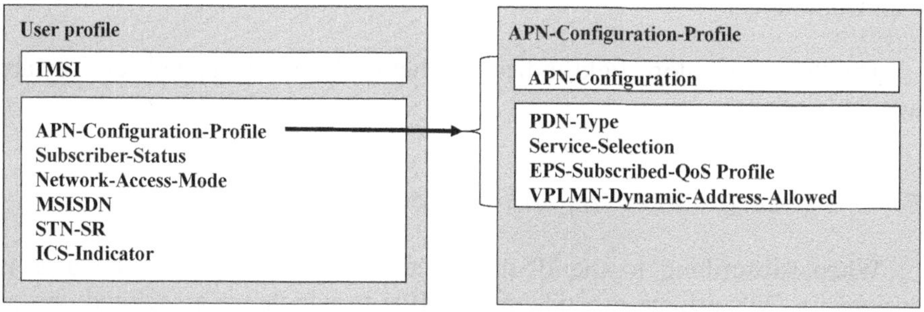

Figure 5.1. *Subscription data to EPS network*

The home subscriber server (HSS) stores the service profile data that is transmitted to the mobility management entity (MME) when attaching the mobile or when changing the profile service.

The "PDP-Type" field indicates the version of IP (Internet Protocol): IPv4, IPv6, IPv4 or IPv6, IPv4 and IPv6.

The “service-selection” field contains the access point name (APN): `ims` in the case of voice or conversational video service, `sos` in the case of emergency call.

The “EPS-subscribed-QoS profile” field indicates the value of the QoS class identifier (QCI) and the allocation and retention priority (ARP).

The “VPLMN-dynamic-address-allowed” field indicates whether roaming is allowed or not.

The “subscriber-status” field indicates whether the telephone service is allowed or not.

The “network-access-mode” field indicates whether the telephone service is available in packet-switched (PS) and circuit-switched (CS) mode or in PS mode only.

The “MSISDN” field contains the mobile subscriber number.

The “STN-SR” field contains the session transfer number for SRVCC, used during the PS-CS inter-system handover, in the case of the implementation of the e-SRVCC function (enhanced Single Radio Voice Call Continuity).

The “ICS-indicator” field indicates whether IMS centralized services are available or not.

5.1.2. *Subscription to the IMS network*

When subscribing to the IP multimedia subsystem (IMS), the user is assigned a IMS private user identity (IMPI) to which are associated one or more service profiles (Figure 5.2).

The HSS entity stores data relating to a user’s service profile which is transmitted to the serving call session control function (S-CSCF) when the registration of the user or when a change of the service profile.

Each service profile includes one or more IMS public user identity (IMPU) in the form of an SIP URI or TEL URI and a service logic in the form of initial filter criteria (iFC) (Figure 5.2).

The field "core network service authorization" of the service profile structure contains an integer number representing the type of media (audio, video) that the user has the right to negotiate in the session description protocol (SDP) message (Figure 5.2).

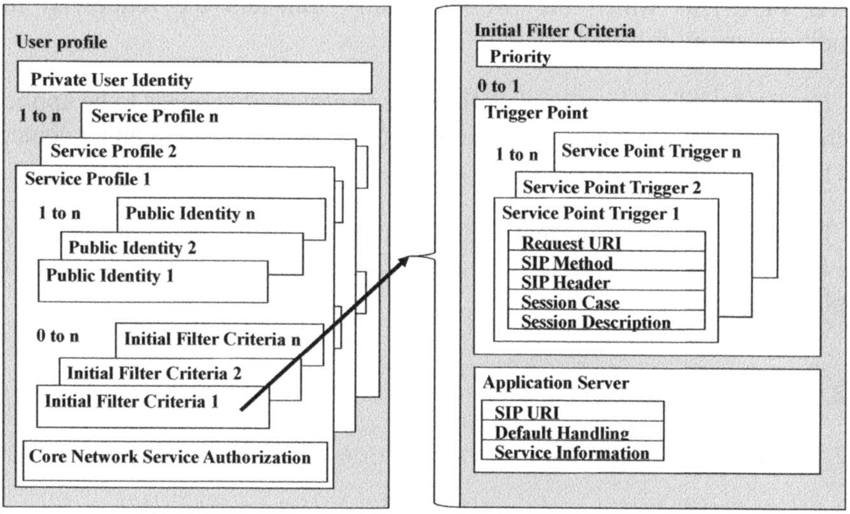

Figure 5.2. *Subscription data to the IMS network*

The first field of iFC data is priority, which determines the order in which the criteria need to be evaluated.

The next field relates to the trigger point, which is a collection of filters (service point triggers)

Each filter consists of a logical operation carried out on the basis of the following parameters:

– Request-URI: the value of the uniform resource identifier (URI) of the request;

– SIP Method: the type of method of the request;

– SIP Header: the content of the headers;

– Session Case: the type of call (outgoing or incoming), for a registered or unregistered mobile;

– Session Description: the content of the fields of the SDP message.

The iFC data also contain the public identity of the telephony application server (TAS) to which the S-CSCF entity transfers the request if the conditions are all fulfilled.

The field "default handling" defines the treatment that needs to be applied to the request (continuation or cessation) if the S-CSCF entity cannot contact the telephony application server.

The field "service information" contains data which are transparent for the HSS and S-CSCF entities. These data are handled only by the telephony application server.

5.2. VoLTE profile service

5.2.1. *Supplementary telephone services*

The user agent (UA) indicates that it supports supplementary telephone service by adding in the REGISTER and INVITE requests, the parameter `+g.3gpp.icsi-ref` associated with the header `Contact`.

```
...
Contact: <sip:192.0.2.101>;expires=600000;+g.3gpp.icsi-
ref="urn:urn-7:3A3gpp-service-ims.icsi.mmtel"
...
```

Table 5.1 lists the supplementary telephone service. The shaded boxes represent supplementary telephone service described in this chapter.

5.2.1.1. *Communication diversion*

CDIV consists of diverting a call (from Alice to Bob) to a different destination (Carol). In this scenario, CDIV is implemented by the telephony application server associated with the callee's (i.e. Bob's) S-CSCF entity.

CDIV (*Communication DIVersion*)
OIP (*Originating Identification Presentation*)
OIR (*Originating Identification Restriction*)
TIP (*Terminating Identification Presentation*)
TIR (*Terminating Identification Restriction*)
MWI (*Message Waiting Indication*)
HOLD (*Communication hold*)
CONF (*CONFerence calling*)
CW (Communication Waiting)
CB (*Communication Barring*)
ECT (*Explicit Communication Transfer*)
MCID (*Malicious Communication Identification*)
CCBS (*Completion of Communications to Busy Subscriber*)
CCNR (*Completion of Communications on No Reply*)
AOC (*Advice Of Charge*)
CUG (*Closed User Group*)
CAT (*Customised Alerting Tone*)
CRS (*Customised Ringing Signal*)
FA (*Flexible Alerting*)

Table 5.1. *Supplementary telephone service*

5.2.1.1.1. CFU

Communication forwarding unconditional (CFU) consists of systematically forwarding an incoming call to another destination. However, the outgoing calls initiated by Bob's UA entity are not affected by this service.

Optionally, a subscription service may be offered to Bob, giving him an indication that CFU is active. This information is transmitted to him when Bob initializes a communication.

The different exchanges relating to CFU are shown in Figure 5.3.

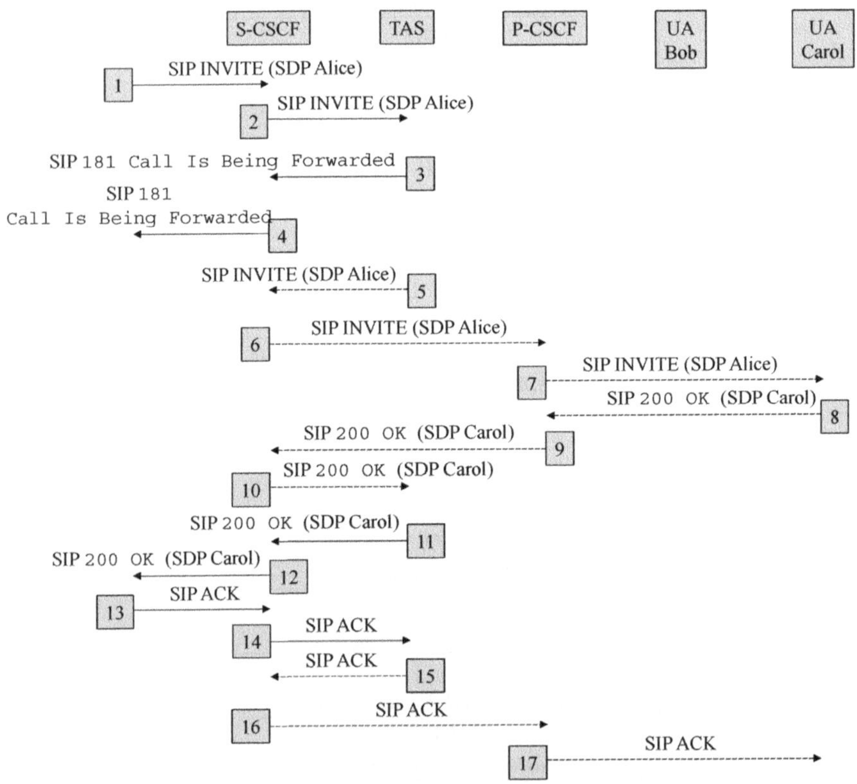

Figure 5.3. *CFU*

1) The INVITE request transmitted by Alice's UA entity is received by Bob's S-CSCF entity.

The INVITE message contains the SDP message of Alice's UA entity.

2) As the result of the iFC data handling is positive, the request is transmitted to the telephony application server.

3), 4) The telephony application server responds to Alice's UA entity with the `181 Call Is Being Forwarded`. This message alerts Alice's UA entity that the call has been transferred.

5), 6), 7) The telephony application server generates an INVITE request, sent to Carol's UA entity. This request includes the SDP message from the INVITE request received from Alice's UA entity.

8), 9), 10) Carol's UA entity responds to the telephony application server with the message `200 OK`.

The response contains the SDP message of Carol's UA entity.

11), 12) The telephony application server transfers the SDP message from Carol's UA entity in the message `200 OK` transmitted to Alice's UA entity.

13), 14) The ACK request from Alice's UA entity acknowledges receipt of the message `200 OK` from the telephony application server.

15), 16), 17) The ACK request from the telephony application server acknowledges receipt of the message `200 OK` from Carol's UA entity.

The media flow is then established between Alice's and Carol's UA entities.

5.2.1.1.2. CFB

Communication forwarding on busy user (CFB) consists of forwarding an incoming call when the callee (Bob) is busy.

The different exchanges relating to CFB are shown in Figure 5.4.

1) to 5) The INVITE request transmitted by Alice's UA entity to Bob's UA entity passes through the telephony application server associated with Bob's S-CSCF entity.

6), 7), 8) Bob's UA entity responds with the message `486 Busy Here`.

This message is intercepted by the telephony application server, which implements communication forwarding to Carol's UA entity.

The following exchanges are identical to those shown for CFU (3 to 17).

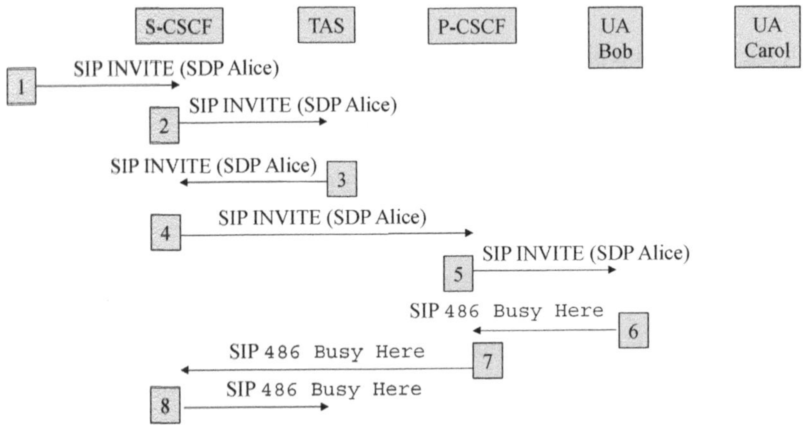

Figure 5.4. *CFB*

5.2.1.1.3. CFNR

Communication forwarding on no reply (CFNR) involves forwarding an incoming call in the case of the lack of a response from the callee (Bob).

The different exchanges relating to CFNR are shown in Figure 5.5.

1) to 5) The INVITE request submitted by Alice's UA entity is received by Bob's UA entity.

6) to 10) On receiving the INVITE request, Bob's UA entity responds with the message `180 Ringing`. This message passes through the telephony application server which starts a timer.

11), 12) At the expiration of the timer, the TAS entity responds to Alice's UA entity with the message `181 Call Is Being Forwarded`. This message can alert Alice's UA entity that her call was being transferred.

13), 14), 15) The TAS entity stops the ringing by sending the CANCEL request to Bob's UA entity.

16), 17), 18) Bob's UA entity responds with the `200 OK` message to the CANCEL request.

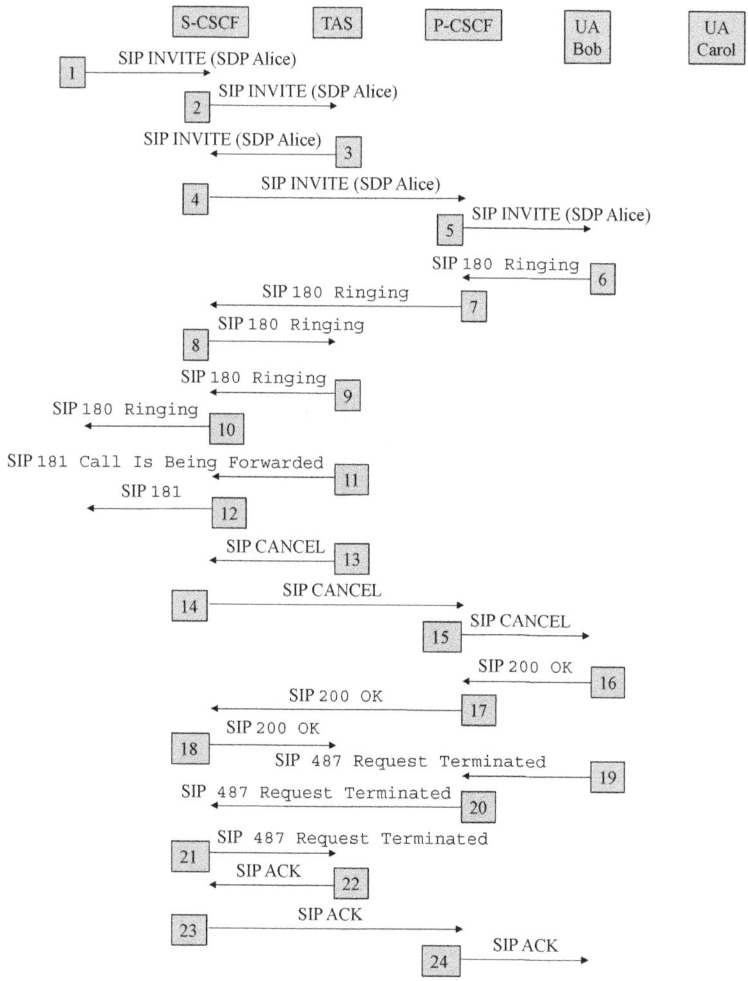

Figure 5.5. *CFNR*

19), 20), 21) Bob's UA entity responds with the message `487 Request Terminated` to the INVITE request.

22), 23), 24) The ACK request from the telephony application server acknowledges the receipt of the message `487 Request Terminated` from Carol's UA entity.

The following exchanges are identical to those described for CFU (5 to 17).

5.2.1.1.4. CFNL

Communication forwarding on not logged-in (CNFL) consists of diverting an incoming call if the callee (Bob) is not logged in with the S-CSCF entity.

The different exchanges relating to CFNL are identical to those for CFU.

5.2.1.1.5. CD

Communication deflection enables the callee (Bob) to deflect a call if he does not want to answer the caller (Alice).

The different exchanges relating to CD are shown in Figure 5.6.

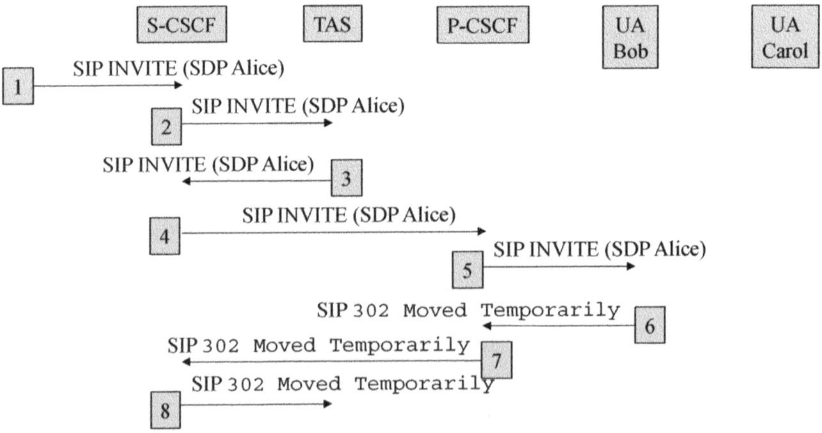

Figure 5.6. *CD*

1) to 5) The INVITE request submitted by Alice's UA entity is received by Bob's UA entity.

6), 7), 8) Bob's UA entity responds with the message `302 Moved Temporarily`. The response contains Carol's URI, to which the call should be forwarded.

The response is intercepted by the TAS entity, which diverts the communication to Carol's UA entity.

The following exchanges are identical to those described for CFU (5 to 17).

5.2.1.2. *Identification presentation*

5.2.1.2.1. OIP and OIR

Originating identification presentation (OIP) consists of presenting the callee (Bob) with the identification of the caller (Alice), asserted by the IMS network.

Alice's UA inserts into the INVITE request the header `P-Preferred-Identity`, containing the public identity IMPU.

```
INVITE sip:bob@homeB.net SIP/2.0
...
P-Preferred-Identity:<sip:alice@home1.net>
...
```

The P-CSCF (Proxy-CSCF) entity replaces this header with the header `P-Asserted-Identity`. The value of this header is presented to Bob's UA entity if the identity of Alice was registered. Otherwise, the P-CSCF entity inserts in the header `P-Asserted-Identity` Alice's identity registered by default.

```
INVITE sip:bob@homeB.net SIP/2.0
...
P-Asserted-Identity:<sip:alice@home1.net>
...
```

Originating identification restriction (OIR) consists of hiding Alice's identification from the callee (Bob).

Alice's UA entity can temporarily mask the presentation of her identity. It has to perform the following operations:

– it has to fill in the header From with the following value: <sip:anonymous@anonymous.invalid>;

– it has to indicate the value id in the header Privacy so that the identity present in the header P-Asserted-Identity is not presented to Bob.

```
INVITE sip:bob@homeB.net SIP/2.0
...
From: "anonymous"
<sip:anonymous@anonymous.invalid>;tag=xyz
P-Asserted-Identity:<sip:alice@home1.net>
Privacy:id
...
```

If Alice has subscribed to OIR, her telephony application server has to insert the value id into the header Privacy. It also has to modify the value of the header From to erase Alice's identity.

If Alice has subscribed to OIR, Alice's UA entity can raise the restriction temporarily by inserting the value none in the header Privacy.

If Bob has not subscribed to OIP or if the header Privacy has the value id, his telephony application server has to remove the header P-Asserted-Identity, modify the value of the header From and remove Alice's identity.

5.2.1.2.2. TIP and TIR

Terminating identification presentation (TIP) consists of showing Alice, in the response 183 Session Progress, the identity of the callee (Bob or Carol in the case of communication forwarding), certified by the IMS network.

```
SIP/2.0 183 Session Progress
...
P-Asserted-Identity: <sip:bob@homeB.net>
...
```

If Alice has not subscribed to TIP or if the messages received contain the header `Privacy` with the value `id`, her telephony application server has to remove the header `P-Asserted-Identity` and modify the value of the header `From`, deleting Bob's or Carol's identity.

Terminating identification restriction (TIR) consists of hiding Bob or Carol's identity from Alice.

The callee's UA entity (i.e. Bob's or Carol's) can temporarily mask the presentation of his/her identity by adding the value `id` to the header `Privacy` in the response.

If the callee (Bob or Carol) has subscribed to TIR, their telephony application server has to insert the header `Privacy` with the value `id`.

If the callee (Bob or Carol) has subscribed to TIR, they can temporarily raise the restriction by indicating the value `none` in the header `Privacy`.

5.2.1.3. *Message waiting indication*

Message waiting indication (MWI) enables the telephony application server to indicate that a communication has occurred and that a voicemail (for instance) has been recorded. In this case, the TAS entity fulfills the function of a voicemail messaging service.

An MWI is transmitted if the user has subscribed to the MWI service.

Alice's UA entity transmits a SUBSCRIBE request to subscribe to the MWI service. This message contains the header `event` with the value `message-summary`. It indicates in the header `Accept` the type of message body (value `application/simple-message-summary`).

```
SUBSCRIBE sip:Alice@homeA.net SIP/2.0
...
Event: message-summary
Accept: application/simple-message-summary
...
```

The request is transmitted to the telephony application server by the S-CSCF entity following the iFC data processing.

The telephony application server verifies that the subscription for Alice's identity, contained in the header `P-Asserted-Identity`, is authorized and responds with `200 OK`. Otherwise, the response is `403 Forbidden`.

When the subscription has been successful, the telephony application server immediately sends a NOTIFY request to synchronize the current state of the messages waiting. This initial notification contains only brief information about the recorded messages.

```
NOTIFY sip:Alice@homeA.net SIP/2.0
...
Subscription-State: active; expires=86399
Event: message-summary
...
Content-Type: application/simple-message-summary
Content-Length: (...)

Messages-Waiting: yes
Message-Account: sip:Alice@homeA.net
Voice-Message: 2/1 (1/0)
Video-Message: 0/1 (0/0)
```

The message body of the NOTIFY request contains the following fields:

– `Message-Account`: this field includes Alice's public identity;

– `Voice-Message 2/1 (1/0)`: this field shows the number of voice messages recorded:

- `2/1`: two new and one old voice messages are recorded,

- `(1/0)`: one of the two new messages is urgent;

– `Video-Message: 0/1 (0/0)`: this field shows the number of video messages recorded:

- `0/1`: one old message is recorded,

- `(0/0)`: no urgent messages are recorded.

If new messages are recorded, the telephony application server sends a new NOTIFY request whose message body provides a detailed description (the caller's identity, the subject of the message, the date) of the new messages.

5.2.1.4. *HOLD*

The HOLD service enables a user to suspend an active communication and come back to it later on.

The different exchanges relating to call parking are shown in Figure 5.7.

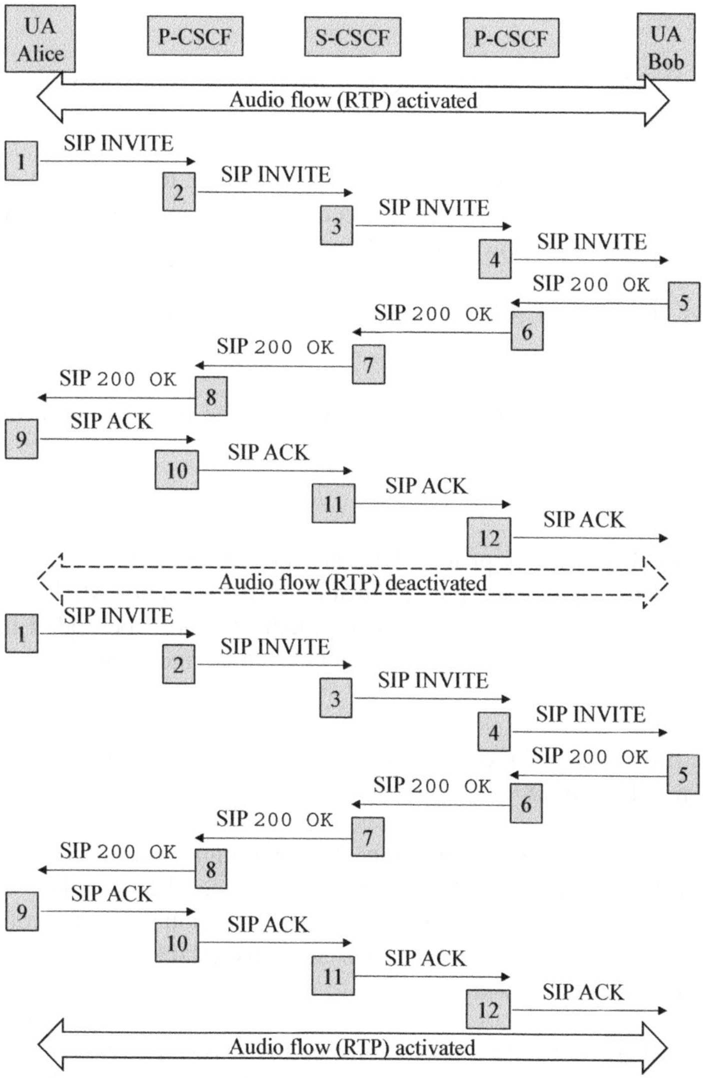

Figure 5.7. *HOLD*

1) to 4) Alice's UA entity sends a re-INVITE request to Bob's UA entity to park the current call. Parking is achieved by modifying the attributes of the audio flow in the SDP message (`a=sendonly`).

5) to 8) Bob's UA entity responds positively with the message `200 OK`. The SDP message is associated with the indication of the attribute (`a=recvonly`).

9) to 12) Alice's UA entity acknowledges the response with the ACK message.

13) to 16) Alice's UA entity sends a re-INVITE to Bob's UA entity to resume the parked communication. Recovery is achieved by modifying the attributes of the media flow in the SDP message (`a=sendrecv`).

17) to 20) Bob's UA entity responds positively with the message `200 OK`.

20) to 24) Alice's UA entity acknowledges the response with the ACK message.

5.2.1.5. *Conference*

The CONF service is provided by an application server comprising the multimedia resource function controller (MRFC) and MRF processor (MFRP). The MRFP entity enables the media flows to be added together. The MRFC entity plays the part of a UA entity and controls the MRFP entity.

The different exchanges relating to establishing the conference service are shown in Figure 5.8.

Two UA entities (Alice and Bob) are communicating. Alice decides to invite Carol and to activate the conference service. Alice puts Bob on hold, initializes a session with Carol, establishes a session with the conference bridge (MFRP) and transfers the two sessions established with Bob and Carol to the conference bridge.

When Alice's UA entity has established the sessions with Bob, Carol and the conference bridge, she begins the transfer of the session established with Bob to the conference bridge by sending the request REFER to the MRFC entity.

The REFER request includes the header `Refer-to` containing the references of the dialog initialized with Bob's UA entity (Bob's URI, header `Call-Id`, parameters `tag` in the headers `From` and `To`).

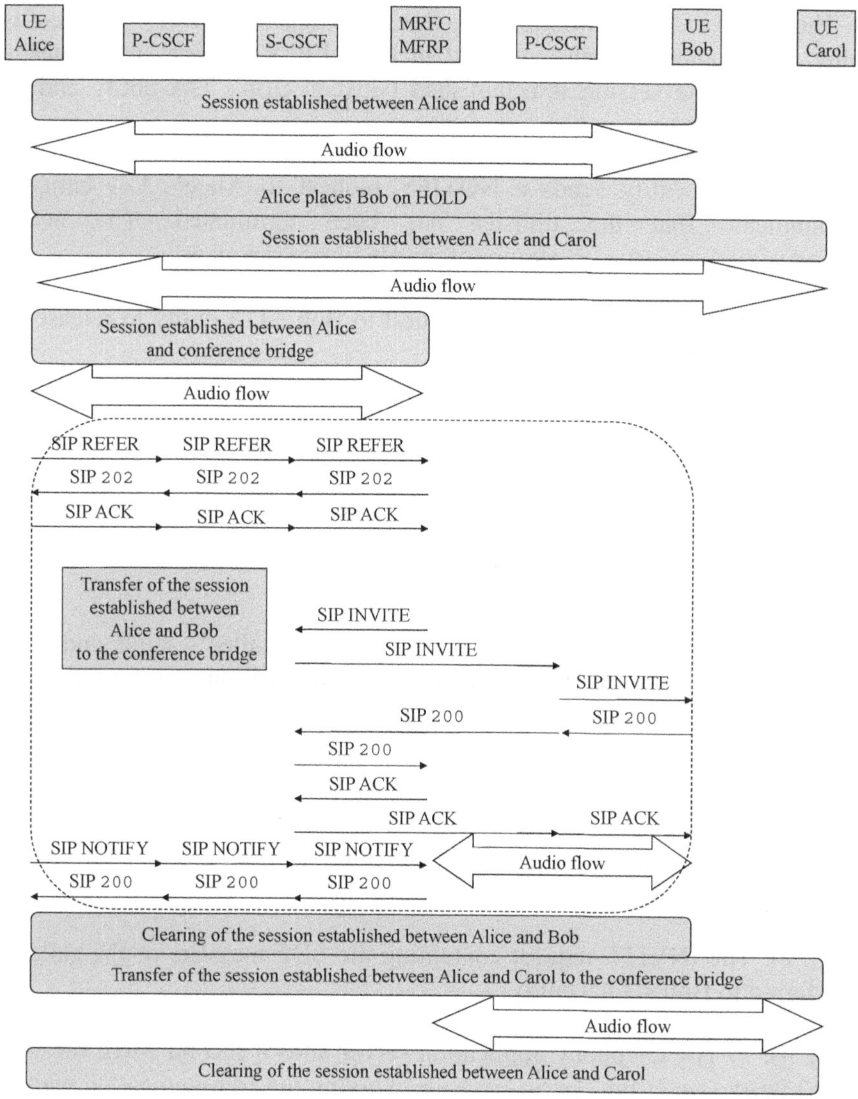

Figure 5.8. *CONF*

The MRFC entity sends a NOTIFY request to Alice's UA entity to indicate that the transfer has been activated. The header `Subscription-State` assumes the value `active`.

The MRFC entity initializes an INVITE request to Bob's UA entity, using the parameters communicated in the REFER request in the header `Replaces`. The session is established between Bob's UA entity and the conference bridge.

The MRFC entity sends a NOTIFY request to Alice's UA entity to communicate that the transfer has been terminated. The header `Subscription-State` assumes the value `terminated`.

Alice's UA entity sends a BYE request to Bob's UA entity to release the previously-established session.

5.2.1.6. *Communication waiting*

Communication waiting (CW) gives information to the caller telling them that the callee is on a different communication but that he has been notified of the call.

Two possible scenarios may occur:

– if the telephony application server knows the callee's status and if the service is activated (case 1), it inserts the communication waiting indication into the INVITE request sent to the callee;

– otherwise (case 2), the TAS entity is informed of the situation in the response given by the callee.

The different exchanges relating to communication waiting, for case 1, are shown in Figure 5.9.

1), 2), The INVITE request containing the SDP message of the caller is transferred to Bob's TAS entity.

3), 4), 5) The telephony application server adds a second XML message (`application/vnd.3gpp.cw+xml`) giving the communication waiting indication, to the message body containing an SDP message (`application/sdp`) in the INVITE request, received from the S-CSCF entity.

The telephony application server indicates in the header `Content-Type` that multiple messages appear in the message body (value `multipart/mixed`).

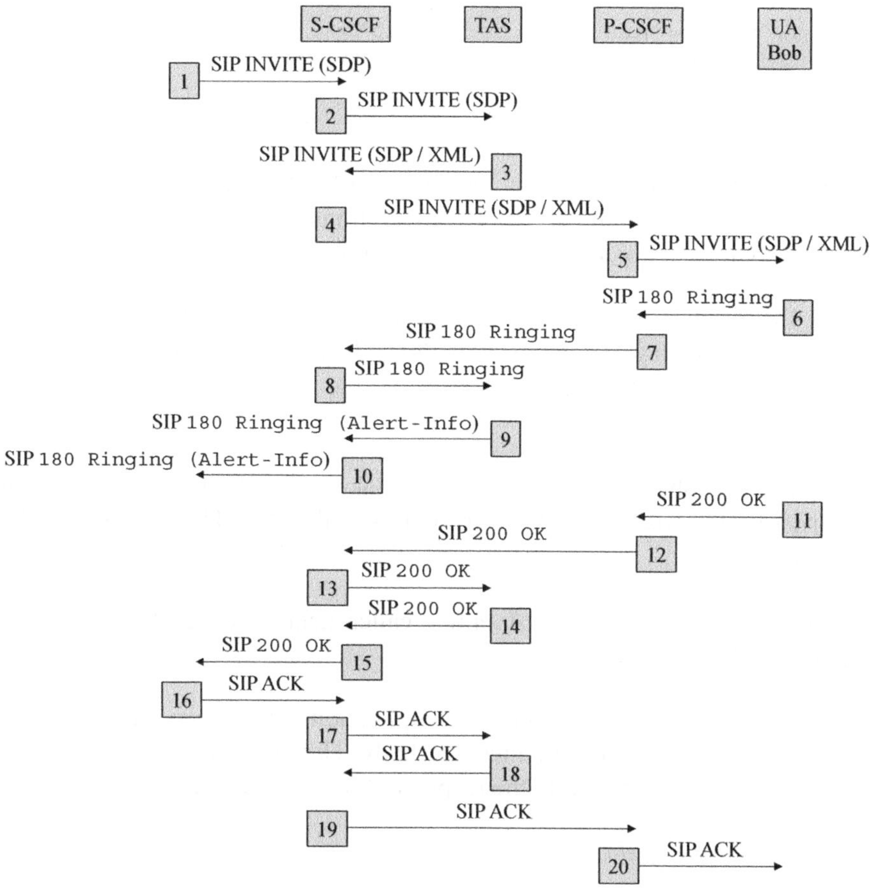

Figure 5.9. *CW*

The telephony application server also includes in this header a parameter `boundary` corresponding to the boundary between the messages (value `boundary1`).

```
-boundary1
Content-Type: application/sdp
(SDP message)
-boundary1
Content-Type: application/vnd.3gpp.cw+xml
<?xml version="1.0"?>
<ims-cw xmlns="urn:3gpp:ns:cw:1.0">
      <communication-waiting-indication/>
</ims-cw>
-boundary1–
```

6), 7), 8) On receiving the message `180 Ringing`, the telephony application server implements the CFNR, to transfer the call to a third party, in case Bob does not pick the waiting communication up.

9), 10) To the message `180 Ringing` received from the P-CSCF entity, the TAS may optionally add the indication to the caller of a call waiting in the header `Alert-Info` (value `urn:alert:service:call-waiting`).

11) to 15) Bob's UA entity decides to take the call and end the current call or park his correspondent, and responds to the INVITE request with the message `200 OK`.

16) to 20) The caller's UA entity caller acknowledges the `200 OK` response with the ACK message.

In the second case, the telephony application server relays the INVITE request, without enriching the message body. However, as before, it inserts the header `Alert-Info` into the response `180 Ringing`.

5.2.1.7. *Communication rejection*

5.2.1.7.1. ICB

Incoming communication barring (ICB) enables the callee (Bob) to reject an incoming call belonging to a certain category of communication. The rule may include the URI of the caller, present in the header `P-Asserted-Identity` or `Referred-By` (in the case of call forwarding).

Bob's telephony application server responds to Alice's UA entity with the message `603 Decline`.

5.2.1.7.2. OCB

Outgoing communication barring (OCB) is able to block an outgoing communication belonging to a certain category of communication.

Bob's telephony application server responds to Alice's UA entity with the message `603 Decline`.

5.2.1.7.3. ACR

Anonymous communication rejection (ACR) enables Bob to reject an anonymous communication.

Bob's TAS entity rejects the communication because the INVITE request contains the header `Privacy` with the value `id`. It responds to the INVITE request with the message `433 Anonymity Disallowed`.

5.2.2. *Audio flow*

The SDP message contains the types of audio codecs proposed by the caller and the codec selected by the callee, among the following list of codecs:

– adaptative multi-rate (AMR);

– AMR-WB (Wide Band);

– enhanced voice services (EVS).

5.2.2.1. *AMR codec*

The AMR codec provides the analog/digital conversion of an audio signal in the frequency band 300–3400 Hz.

The AMR codec produces 20 ms frames and the resulting flow can have the following values: 12.2 kbps, 10.2 kbps, 7.95 kbps, 7.40 kbps, 6.70 kbps, 5.90 kbps, 5.15 kbps and 4.75 kbps.

The AMR codec uses discontinuous transmission (DTX) associated with voice activity detection (VAD) and comfort noise generation (CNG) to reduce the flow rate during periods of silence.

The SDP offer contains the values of the payload type field of the real-time transport protocol (RTP) header that identifies the codec type used.

```
...
m=audio 49152 RTP/AVP 97 98
a=rtpmap:97 AMR/8000/1
a=rtpmap:98 AMR/8000/1
a=fmtp:98 octet-align=1
...
```

The value 97 corresponds to the bandwidth-efficient format, for which only the payload is aligned on byte boundaries, fewer padding bits being thus added.

The value 98 corresponds to byte-aligned format, where all fields in the payload of the RTP segment are aligned individually on byte boundaries (`a = fmtp: 98 byte-align = 1`).

The value of 97 for the payload type field is listed first as the corresponding format has priority.

5.2.2.2. *AMR-WB codec*

The AMR-WB codec has the same technical features as the AMR codec and improves voice quality by increasing the bandwidth (50–7000 Hz) for the audio signal.

Several configurations (A, B, C) and several rates per configuration are defined:

– configuration A: 6.6 kbps, 8.85 kbps and 12.65 kbps;

– configuration B: 6.6 kbps, 8.85 kbps, 12.65 kbps and 15.85 kbps;

– configuration C: 6.6 kbps, 8.85 kbps, 12.65 kbps and 23.85 kbps.

The SDP offer is more important because it must include the AMR and AMR-WB codecs and can propose, for each type of codec, both formats:

– the values 97 and 99 of the Payload Type field correspond to the bandwidth-efficient format for AMR-WB and AMR codecs;

– the values 98 and 100 of the payload type field correspond to the byte-aligned format for AMR-WB and AMR codecs.

The values 97 and 98 of the payload type field are listed first because the AMR-WB codec has priority over the AMR codec.

```
...
m=audio 49152 RTP/AVP 97 98 99 100
a=rtpmap:97 AMR-WB/16000/1
a=rtpmap:98 AMR-WB/16000/1
a=fmtp:98 octet-align=1
a=rtpmap:99 AMR/8000/1
a=rtpmap:100 AMR/8000/1
a=fmtp:100 octet-align=1
...
```

5.2.2.3. *EVS codec*

The EVS concept is a new family of codecs whose different types depend on the bandwidth of the audio signal:

– Narrow band (NB), for an audio signal whose bandwidth is 300–3400 Hz;

– Wide band (WB), for an audio signal whose bandwidth is 50–7000 Hz;

– Super wide band (SWB), for an audio signal whose bandwidth is 50–14000 Hz;

– Full band (FB), for an audio signal whose bandwidth is 20–20000 Hz.

The flow rates obtained have the following values: 5.9 kbps, 7.2 kbps, 8 kbps, 9.6 kbps, 13.2 kbps, 16.4 kbps, 24.4 kbps, 32 kbps, 48 kbps, 64 kbps, 96 kbps and 128 kbps.

The rate for the NB type is in the range between 5.9 and 24.4 kbps.

The rate for the WB type is in the range between 5.9 and 128 kbps.

The rate for the SWB type is in the range between 9.6 and 128 kbps.

The rate for the SWB type is in the range between 16.4 and 128 kbps.

The NB-type and WB-type EVS codecs provide the following improvements compared to the AMR and AMR-WB codecs:

– for the same quality, the rate is decreased;

– for achieving equivalent rate, the quality is improved.

The SWB-type and FB-type EVS codecs are reserved for coding music signals.

The SDP offer relating to EVS codec must also include the AMR and AMR-WB codecs, EVS codec having priority.

The rate required for reservation of resources in the 4G network is equal to 42 kbps (field `b = AS: 42`).

The EVS codec is WB-type because the sampling frequency is 16000 Hz. The rate of WB-type EVS codec is in the range between 5.9 and 24.4 kbps.

```
...
m=audio 49152 RTP/AVP 97 98 99 100 101
b=AS:42
a=rtpmap:97 EVS/16000/1
a=fmtp:97 br=5.9-24.4
a=rtpmap:98 AMR-WB/16000/1
a=rtpmap:99 AMR-WB/16000/1
a=fmtp:99 octet-align=1
a=rtpmap:100 AMR/8000/1
a=rtpmap:101 AMR/8000/1
a=fmtp:101 octet-align=1
...
```

5.3. ViLTE profile service

5.3.1. *Supplementary conversational video service*

The supplementary telephone service described for VoLTE service is extended for the ViLTE service, with supplements for communication hold and conference.

In the case of communication hold, the IMS network can take the decision to reduce the bandwidth for audio and video streams.

In the case of conference, the following configurations are possible:

– a UA entity invited to the conference can only enable the audio stream and disable the video stream indicating in the SDP message, a value equal to ZERO for the port number;

– a UA entity participating in a conversational video conference can disable the video stream and keep the audio stream;

– if the UA entity responsible for the conversational video conference removes the video stream, the IMS network may decide to convert conversational video conference in an audio conference for all participants.

5.3.2. *Video flow*

The SDP message contains the types of video codecs proposed by the caller and the codec selected by the callee, among the following codecs list:

– H.264 codec, with constrained baseline profile (CBP), level 1.2;

– H.265 with main profile (MP) level 3.1 main tier.

5.3.2.1. *H.264 codec*

H.264 profile is identified by the parameter `profile-level-id` in the SDP offer.

The maximum image size (number of pixels) and the maximum number of frames per second for the CBP profile level 1.2 can take several values:

– size 176 × 144, 60.6 frame/s;

– size 320 × 240, 20 frame/s;

– size 352 × 288, 15.2 frame/s.

Several actual sizes of the image can be negotiated and are identified by the parameter `a = imageattr` in the SDP offer.

The maximum rate supported by the profile is equal to 384 kbps. The reserved actual reserved is indicated by the parameter `b = AS` in the SDP offer.

```
...
m=video 49154 RTP/AVP 99
b=AS:315
a=rtpmap:99 H264/90000
a=fmtp:99 profile-level-id=42e00c
a=imageattr:99 send [x=176,y=144] [x=224,y=176]
[x=272,y=224] [x=320,y=240] recv [x=176,y=144]
[x=224,y=176] [x=272,y=224] [x=320,y=240]
...
```

5.3.2.2. *H.265 codec*

The H.265 codec presents the following improvements compared to H.264 codec:

– for the same quality, the rate is decreased;

– for an equivalent rate, the quality is improved.

The maximum image size (number of pixels) and the maximum number of frames per second for the MP profile level 3.1 main tier can take several values:

– size 720 × 576, 75 frame/s;

– size 960 × 540, 60 frame/s;

– size 1280 × 720, 33.7 frame/s.

In the following example, the mobile is equipped with a 5-inch screen which supports an 848 × 480 frame size and 25 frames per second.

The SDP offer provides H.265 codec, H.264 level 3.1 for the frame size 848 × 480 and H.264 level 1.2 for the frame size 320 × 240.

The required rate is equal to 690 kbps for a H.264 level 3.1 and 540 kb/s for the H.265 codec. The rate indicated by the parameter `b = AS` is the maximum rate.

```
...
m=video 49154 RTP/AVP 98 97 100 99
b=AS:690
a=rtpmap:100 H264/90000
a=fmtp:100 profile-level-id=42e01f
a=imageattr:100 send [x=848,y=480] recv [x=848,y=480]
a=rtpmap:99 H264/90000
a=fmtp:99 profile-level-id=42e01f
a=imageattr:99 send [x=320,y=240] recv [x=320,y=240]
a=rtpmap:98 H265/90000
a=fmtp:98 profile-id=1; level-id=93
a=imageattr:98 send [x=848,y=480] recv [x=848,y=480]
a=rtpmap:97 H265/90000
a=fmtp:97 profile-id=1; level-id=93
a=imageattr:97 send [x=320,y=240] recv [x=320,y=240]
...
```

6

Interconnections

6.1. Interconnection CS network

6.1.1. *Functional architecture*

The CS (Circuit-Switched) network is the public switched telephone network (PSTN) or public land mobile network (PLMN) implementing time-division multiplexing (TDM).

The functional architecture of the IMS network for interconnection to the CS network is described in Figure 6.1.

The interconnection between the IMS network and the CS network implements new entities:

– breakout gateway control function (BGCF), to determine the entity in charge of interconnection, only for outgoing calls;

– media gateway control function (MGCF), performing the translation of telephone signaling and for controlling the interconnection gateway;

– IMS-MGW (media gateway), adapting the transport formats of traffic flows and signaling flows between the IMS and CS networks.

The BGCF entity processes the INVITE requests sent by the serving call session control function (S-CSCF) in case the session needs to be forwarded to an interconnection. This relates to calls to the users connected to the PSTN or PLMN networks.

From the telephone number, the BGCF entity determines the next hop for the routing of the SIP message. It has to choose the MGCF entity that is responsible for interconnection with the PSTN or PLMN networks.

If the interconnection entity is in a third-party network, the BGCF entity transmits the SIP message to another BGCF entity in that third-party network.

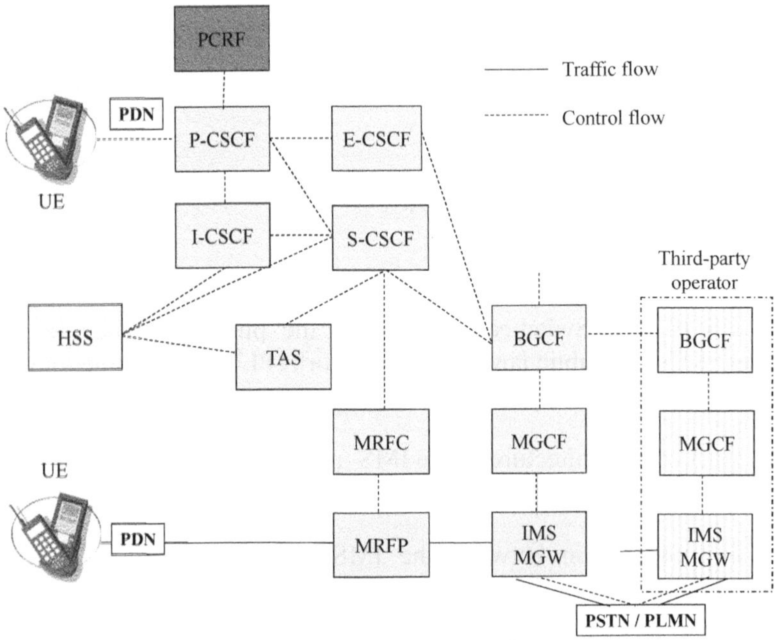

Figure 6.1. *Functional architecture of IMS network interconnection with CS network*

The MGCF entity controls the establishment, the maintenance and the release of the connections in the IMS-MGW entity. A connection represents an association between an end-point related to the interface with the PSTN or PLMN networks and an end-point related to the interface with the IP network.

The MGCF entity performs the translation between SIP (Session Initiation Protocol) signaling used in the IMS network and telephone signaling at the interconnection between IMS or CS networks, using the following protocols:

– ISUP (ISDN User Part), for interconnection in TDM mode, with PSTN and PLMN networks;

– BICC (Bearer Independent Call Control), for interconnection in IP mode with the PLMN networks corresponding to the Release 4;

– SIP-I (SIP with encapsulated ISUP), for interconnection in IP mode with the PLMN networks corresponding to the Release 4.

H.248 signaling is used by the MGCF entity to control the IMS-MGW interconnection gateway, whose deployment is mandatory for the interconnection in TDM mode (case of ISUP signaling) and optional for the interconnection in IP mode (the case of BICC or SIP-I signaling).

The IMS-MGW performs the conversion of protocols relating to the media flows between the two end-points. It contains the treatments carried out on the media flows, such as transcoding (modification of the type of codec between the two end-points), echo cancellation and transmission of tones and announcements.

6.1.2. *Protocol architecture*

6.1.2.1. *ISUP signaling*

The protocol ISUP defines the procedures used to establish, manage and release the circuits which transport the voice signals.

Initial address message (IAM) is transmitted between digital switches to notify the request for establishment of a communication. It contains the telephone numbers of the caller and the callee;

Address complete message (ACM) is returned by the destination switch to indicate that the ringing has been activated for the callee.

Answer message (ANM) is transmitted by the destination switch to indicate that the callee has picked up the telephone.

Release (REL) message is sent to release the resources when a user hangs up the telephone.

Release complete (RLC) message is transmitted to acknowledge the REL message.

6.1.2.2. *BICC signaling*

The BICC protocol is used for signaling exchanged between, on one hand, the MGCF entity of the IMS network, and, on the other hand, the entity of the CS network responsible for interconnection of the telephone signaling.

The BICC protocol was developed from the ISUP protocol that it is similar for basic call procedures and functions of supplementary telephone services basically.

The BICC protocol also implements a mechanism which allows for the exchange of information linked to the support command of the media flow on the IP network between, on one hand, the IMS-MGW interconnection gateway, and, on the other hand, the entity of the CS network responsible for interconnection of the media flow.

This information concerns IP addresses, UDP port numbers, types of media (voice, video and data) and formats used for this media (codecs for voice and video protocols for data).

This information is tunneled in BICC messages containing signaling such as for example the IAM message for communication establishment or in the BICC APM (Application Transport Mechanism) message which does not transport any signaling information.

The IP bearer control protocol (IP BCP) protocol is a control protocol for media flow which aims to ensure the exchange of necessary information to establish or modify media flow characteristics.

The IPBCP protocol is initialized by the IMS-MGW and passed to the MGCF entity thanks to the H.248 protocol, for which a specific package is defined. The package defines additional properties which can appear in the terminations and contexts.

The IPBCP protocol defines four types of message:

– REQUEST message is sent by the MGCF entity to launch a request to establish or modify a media flow on the IP network;

– ACCEPTED message is sent by the MGCF entity which receives a message of media flow establishment or modification, if it accepts the request;

– CONFUSED message is sent by the MGCF entity in response to a media flow modification or establishment if it cannot deal with the received request message;

– REJECTED message is sent by the MGCF entity in response to a request for media flow establishment or modification if it refuses the request.

The bearer control tunnelling protocol (BCTP) protocol is used for the encapsulation of support control protocol for media flow.

The header of the BCTP protocol contains the following indications:

– the version of the BCTP protocol;

– the type of support control protocol for media flow.

6.1.2.3. *SIP-I signaling*

SIP-I signaling allows for the transport of ISUP messages encapsulated by SIP messages.

The ISUP IAM message is carried by the SIP INVITE message.

The ISUP ACM message is carried by the provisional response SIP `180 Ringing` to the SIP INVITE request.

The ISUP ANM message is carried by the SIP `200 OK` response to the SIP INVITE request.

The ISUP REL message is carried by the SIP BYE message.

The ISUP RLC message is carried by the SIP `200 OK` response to the SIP BYE request.

6.1.2.4. *H.248 signaling*

The connection model describes the logical entities contained in the IMS-MGW entity that the MGCF entity can control.

The main abstractions used in this connection model are contexts and terminations.

A termination sends and/or collects several flows. The media flow parameters are encapsulated in the termination.

A context is a package of allocated terminations. The NULL context contains all terminations which are not allocated to another termination, as is the case with idle time-slots on the TDM-mode interface.

H.248 messages have a header which contains the identity of the transmitter and the version of the protocol. Several transactions can be concatenated in a H.248 message. The transactions contained in a message are dealt with independently and no order is predetermined.

Each transaction is indicated by a transaction identifier and consists of one or several actions (Figure 6.2).

An action consists of a series of commands to modify or examine the context property. Each action usually specifies a context identifier (Figure 6.2).

Each command consists of an identifier and a termination descriptor. The descriptor contains the parameters of a termination concerning a command. A descriptor consists of a name and a list of elements (Figure 6.2).

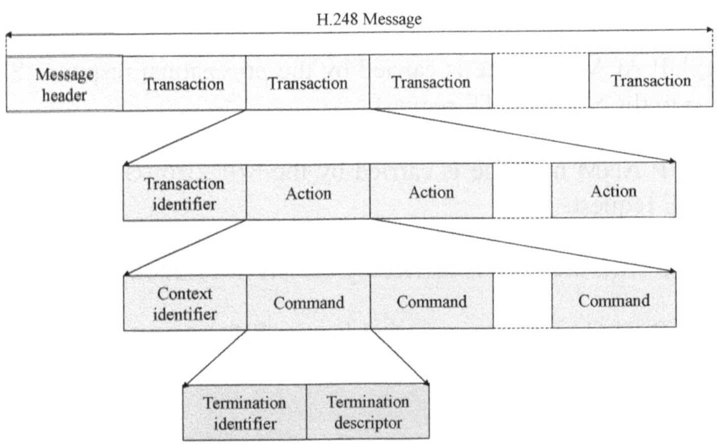

Figure 6.2. *H.248 structure message*

The terminations which are physical entities have a semi-permanent existence like a time-slot in a time-division multiplexing (TDM).

The temporary terminations are real-time transport protocol (RTP) flow, in the case of the interface with the IP network.

The protocol provides commands to manipulate the logical entities, the contexts and the terminations of the connection model.

ADD command adds a termination to a context. Applied to the first termination of a context, it also serves to create a context.

MODIFY command modifies the properties of a termination.

SUBTRACT command disconnects a termination from its context and sends back statistics about the participation of this termination in this context. Applied to the last termination of a context, it serves to cancel this context.

MOVE command moves a termination to another context.

AuditValue command sends back the current states of the properties and statistics associated with the terminations.

AuditCapabilities command sends back all the possible values of termination properties authorized by the IMS-MGW gateway.

NOTIFY command allows the IMS-MGW gateway to inform the MGCF entity about the development of events in this gateway.

ServiceChange command allows the IMS-MGW gateway to signal to the MGCF entity that a termination or a group of terminations is about to be disabled or has just been enabled.

This command is also used by the IMS-MGW to register with the MGCF entity. The MGCF entity can also use this command to request the IMS-MGW to disable a termination or a group of terminations.

Most commands are reserved for the specific use of the MGCF entity to control the IMS-MGW gateway. The exceptions are NOTIFY and ServiceChange commands, the first being sent by the IMS-MGW and the second being able to be sent by one of the two entities.

6.1.2.5. *Interfaces*

The Mi interface is the reference point between the S-CSCF and BGCF entities. This interface supports SIP and SDP signaling.

The Mj interface is the reference point between the BGCF and MGCF entities. This interface supports SIP and SDP signaling.

The Mg interface is the reference point between the CSCF and MGCF entities. This interface supports SIP and SDP signaling.

The Mn interface is the reference point between the MGCF and IMS-MGW entities. This interface supports H.248 signaling.

The IMS-MGW performs the conversion of the transport protocols relating to the signaling exchanged between the MGCF, depending on signaling transport over IP (SIGTRAN) model, the PSTN and PLMN networks, depending on signaling system 7 (SS7) model (Figure 6.3).

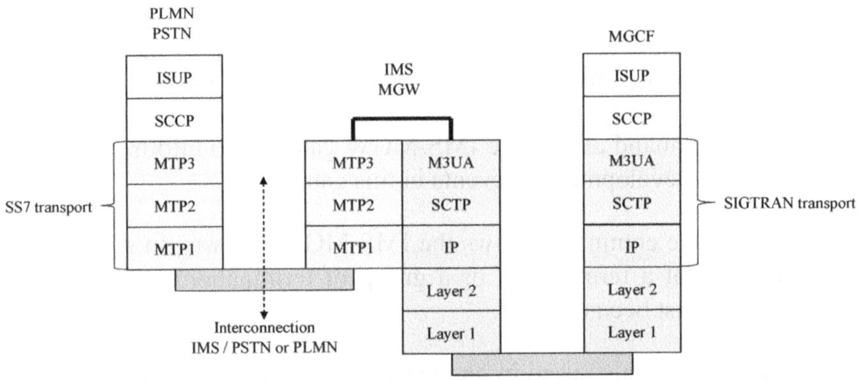

Figure 6.3. *Transport of ISUP signaling*

The IMS-MGW performs, in the case of the ISUP signaling, the conversion of protocols relating to the media flows between the two end-points, corresponding, firstly, to RTP flow, and, secondly, to TDM flow (Figure 6.4).

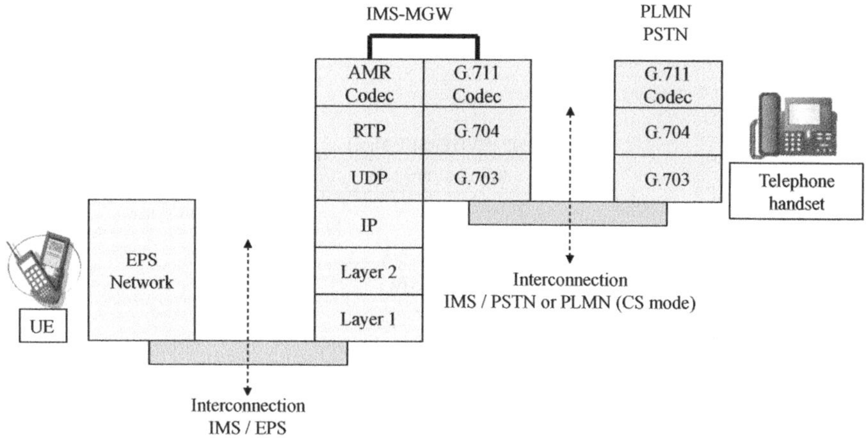

Figure 6.4. *Voice transport*

6.1.3. *Session establishment*

6.1.3.1. *Session establishment initiated by IMS network*

The procedure for establishing the session corresponds to a call generated by the Alice's user agent (UA) connected to the IMS network to Bob's telephone terminal connected to a CS network.

The procedure for establishing the session initiated by the IMS network for ISUP or BICC telephone signaling is illustrated in Figure 6.5.

1) The call is generated by Alice's UA entity, by way of an INVITE request whose identity TEL URI contains the telephone number of the callee.

The S-CSCF entity carries out an ENUM resolution with the DNS server, to convert the TEL URI into a SIP URI.

The INVITE message contains the session description protocol (SDP) message of Alice's UA entity.

As the callee is not a client of the IMS network, the response from the DNS server is negative and the S-CSCF entity then routes the INVITE request to the BGCF entity.

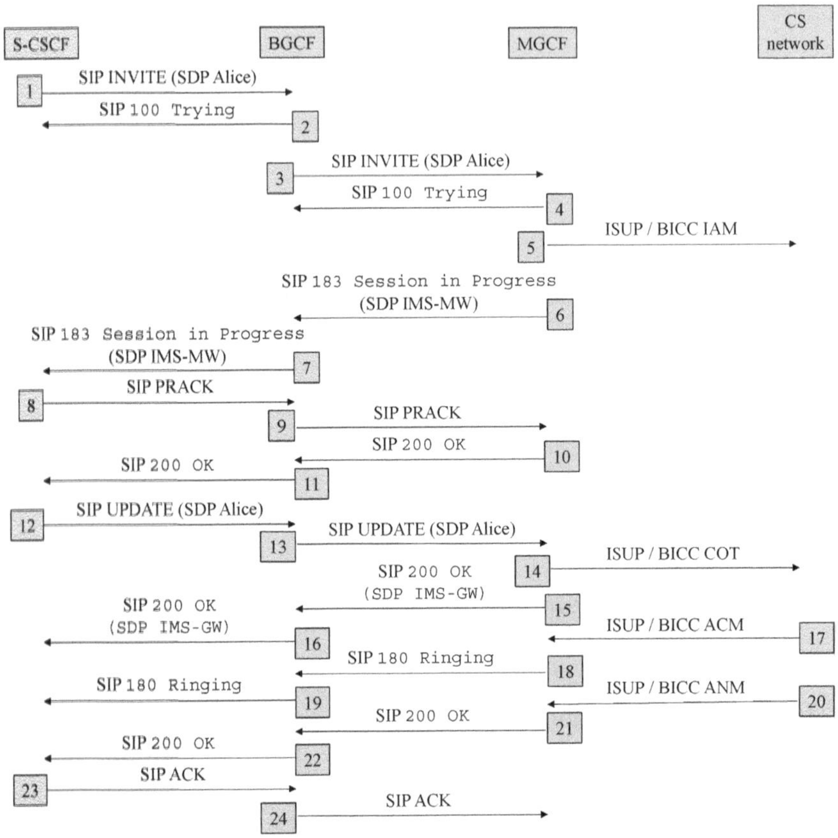

Figure 6.5. *Session establishment initiated by IMS network ISUP or BICC signaling*

2) The BGCF entity responds to the S-CSCF entity with the `100 Trying` message that allows the blocking of the retransmission timer of the INVITE request.

3) The BGCF entity determines the MGCF entity which interconnects with the CS network, either from an ENUM resolution, or from internal correspondence tables, and forwards the INVITE request.

4) The MGCF entity responds to the BGCF entity with the `100 Trying` message that allows the blocking of the retransmission timer of the INVITE request.

5) The MGCF entity converts the INVITE request into an ISUP/BICC IAM message. This message is transmitted to the IMS-MGW using SIGTRAN transport, and then to the network in CS mode.

When the MGCF entity receives the INVITE request, it performs the following operations with the IMS-MGW gateway, in the case of ISUP interconnection:

– it transfers to the IMS-MGW gateway the SDP message associated with the INVITE request;

– it commands the constitution of the context and the end-points at the level of the IMS-MGW gateway;

– it retrieves the characteristics of the media flow of the interface on the IP side and on the TDM side of the IMS-MGW gateway.

In the case of a BICC interconnection, the configuration of the IMS-MGW gateway can for example correspond to a transcoding of voice:

– adaptive multi-rate/narrow band (AMR/NB) codec, at the 4G network side;

– half rate (HR) or full rate (FR) codec, at 2G network side.

6), 7) The MGCF entity transmits the response `183 Session Progress` to the INVITE request in which the SDP messages recovered from the IMS-MGW gateway is inserted.

8), 9) The subsequent PRACK request acknowledges of the provisional response `183 Session Progress`.

10), 11) The `200 OK` message is the response to the PRACK request.

12), 13) The confirmation of the resource reservation on the 4G network is indicated to the MGCF entity in an SDP offer contained in the UPDATE request.

14) The MGCF entity transmits to the CS network the ISUP/BICC COT to inform him that the resource is available on the caller’s side.

15), 16) The `200 OK` message is the response to the UPDATE request.

17) The network operating in CS mode reserves the resources and responds with an ISUP ACM message indicating that the call has been presented to the callee.

18), 19) The MGCF entity transmits the provisional response `180 Ringing` to the INVITE request to trigger the ring back tone at Alice's UA entity.

20) When the callee hangs up, the MGCF entity receives the ISUP/BICC ANM message.

The MGCF entity connects the end-points at the level of the IMS-MGW gateway.

21), 22) The MGCF entity sends the definitive response `200 OK` to Alice's UA entity for stopping the ring back tone.

23), 24) The `200 OK` response is acknowledged by the ACK request.

The procedure for establishing the session initiated by the IMS network for SIP-I signaling is shown in Figure 6.6.

The procedure for establishing the session initiated by the IMS network for SIP-I signaling, takes that described for ISUP/BICC signaling, with the following modifications.

5) The ISUP IAM message is transmitted in a SIP INVITE message that also contains the SDP message of Alice's UA entity.

6) The CS network responds to MGCF entity with the `100 Trying` message which allows the blocking of the retransmission timer of the INVITE request.

7) The provisional response `183 Session in Progress` to the INVITE request is generated by the CS network and contains the SDP message of the gateway of this network.

12) The PRACK request is extended to the CS network.

13) The `200 OK` response to the PRACK request is generated by the CS network.

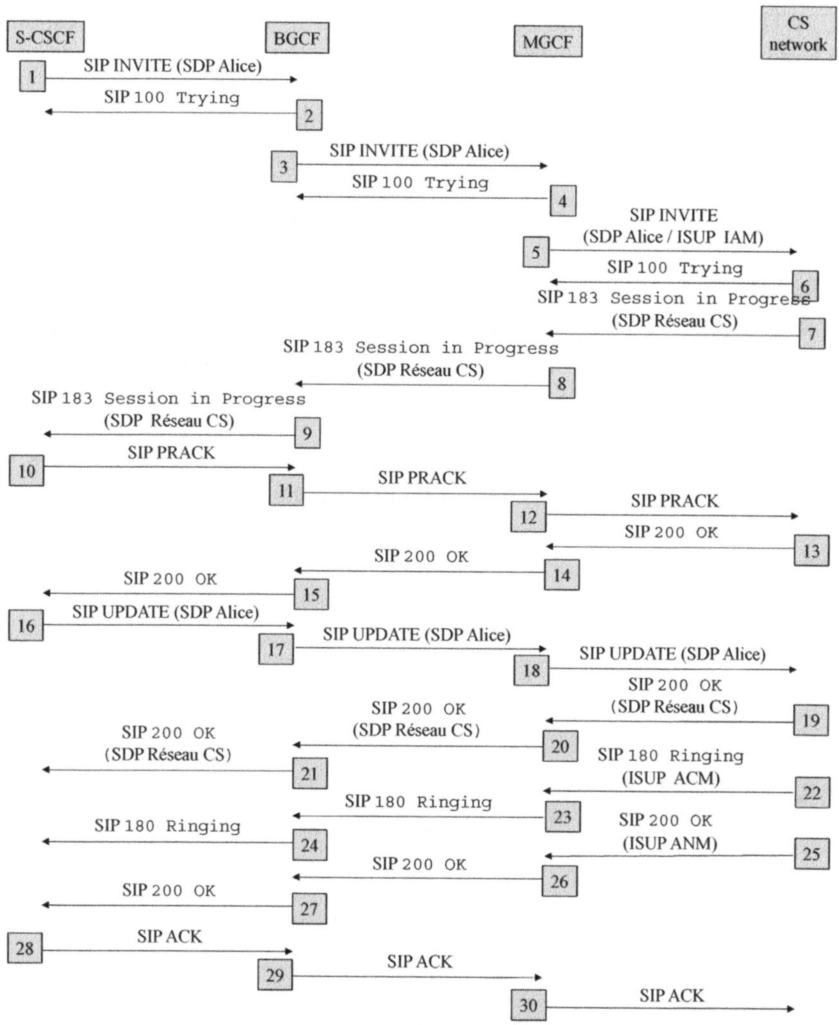

Figure 6.6. *Session establishment initiated by IMS network SIP-I signaling*

18) The UPDATE request is extended to the CS network.

19) The `200 OK` response to the UPDATE request is generated by the CS network.

22) The ISUP ACM message is transmitted in the provisional response 180 Ringing to the INVITE request.

25) The ISUP ANM message is transmitted in the `200 OK` response to the INVITE request.

30) The IP ACK request is extended to the CS network.

6.1.3.2. *Session establishment initiated by CS network*

The procedure for establishing the session corresponds to a call initiated by the telephone terminal of Alice connected to the CS network to Bob's UA entity connected to the IMS network.

The procedure for establishing the session initiated by the CS network for the BICC or ISUP signaling is described in Figure 6.7.

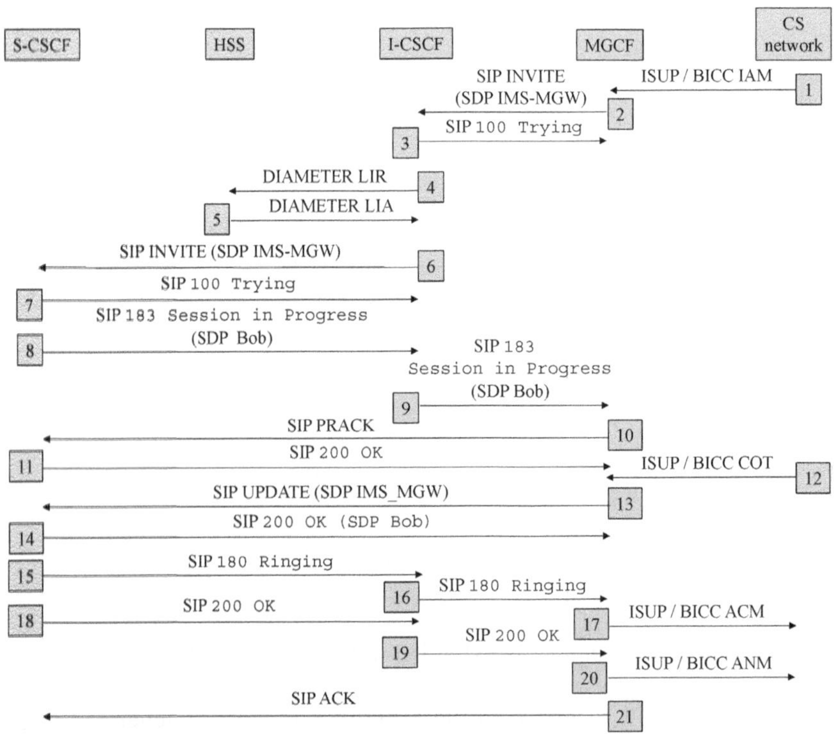

Figure 6.7. *Session establishment initiated by CS network ISUP or BICC signaling*

1) The call is generated by the CS network, and results, at the level of the MGCF entity, in the receipt of the message ISUP/BICC IAM.

The MGCF entity creates the end-points with the IMS-MGW gateway and retrieves the message, describing the characteristics of the termination of the IMS-MGW gateway on the IP side.

2) The MGCF entity generates an INVITE request with the identity TEL URI containing the telephone number contained in the message ISUP/BICC IAM, associating the SDP message given by the IMS-MGW gateway and forwards the INVITE message to the Interrogating-CSCF (I-CSCF).

The MGCF entity shall indicate in the SDP message that the preconditions for the resource reservation are needed.

3) The I-CSCF entity responds to the MGCF entity with the `100 Trying` message that allows the blocking of the retransmission timer of the INVITE request.

4) The I-CSCF entity sends to the home subscriber server (HSS) the DIAMETER LIR (Location-Information-Request) message to retrieve the IP address of the S-CSCF entity which registered Bob's UA entity.

5) The HSS entity provides the IP address of the S-CSCF entity in the DIAMETER LIA (Location-Information-Answer) message.

6) The I-CSCF entity forwards the SIP INVITE request to the S-CSCF entity.

7) The S-CSCF entity responds to the I-CSCF entity with the `100 Trying` message that allows the blocking of the retransmission timer of the INVITE request.

8), 9) On receipt of the message `183 Session in Progress`, the MGCF entity transfers the SDP message from the Bob's UA entity to the IMS-MGW gateway, thereby supplementing the characteristics of the end-point on the IP network.

10) The PRACK request acknowledges the provisional response `183 Session in Progress`.

11) The `200 OK` message is the response to the PRACK request.

12) The message ISUP/BICC COT tells the MGCF entity that the resource has been reserved on the CS network.

13) The confirmation of the resource reservation on the CS network is indicated to Bob's UA entity in an SDP offer contained in the UPDATE request.

14) The confirmation of the resource reservation on the 4G network is indicated to the MGCF entity in an SDP offer contained in the `200 OK` response to the INVITE request.

15), 16) When Bob's UA entity has confirmation of the resource reservation in the 4G network, the telephone rings and the MGCF entity receives the message `180 Ringing`.

17) The MGCF entity generates the message ISUP / BICC ACM, sent to the CS network.

18), 19) When Bob picks up, the MGCF entity receives the message `200 OK` and connects the end-points of the IMS MGW.

20) The MGCF entity transmits the message ISUP / BICC ANM to the network in CS mode.

21) The `200 OK` response is acknowledged by the ACK request.

The procedure for establishing the session generated by the CS network for SIP-I signaling is described in Figure 6.8.

The procedure for establishing the session generated by the CS network for SIP-I signaling, takes that described for ISUP/BICC signaling, with the following modifications.

1) The ISUP IAM message is transmitted in the INVITE message which also contains the SDP message of the CS network gateway.

2) The MGCF entity responds to the CS network with the `100 Trying` message that allows the blocking of the retransmission timer of the INVITE request.

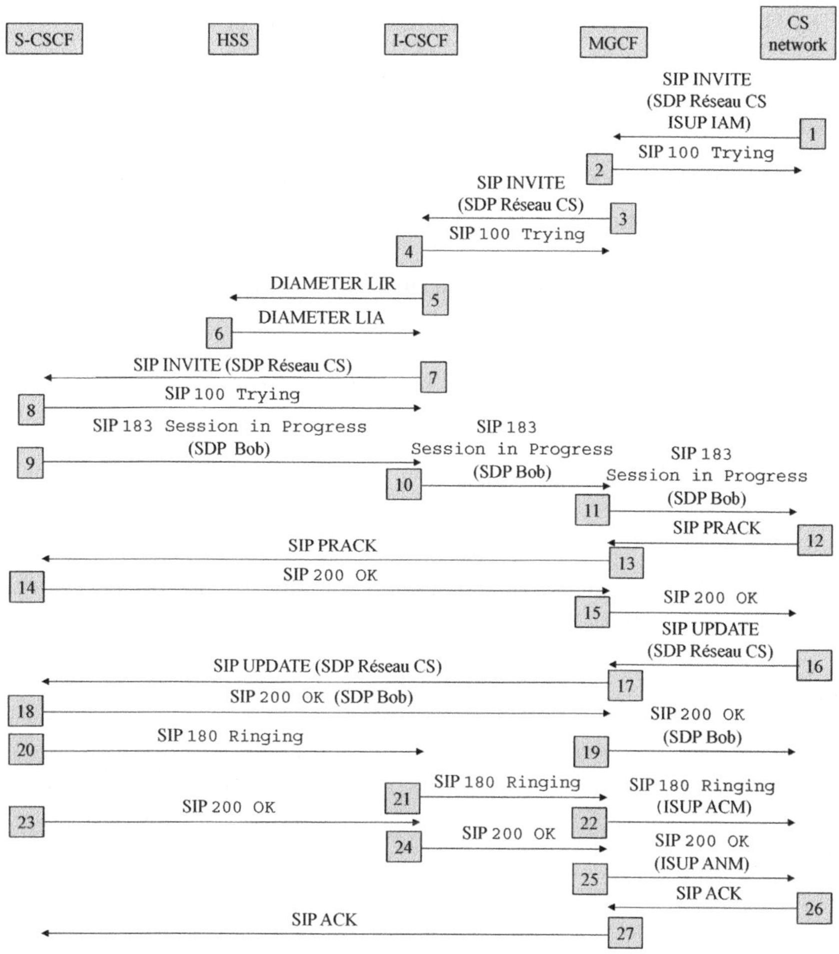

Figure 6.8. *Session establishment initiated by CS network SIP-I signaling*

11) The provisional response `183 Session in Progress` to the INVITE request is extended to the CS network.

12) The PRACK request is generated by the CS network.

15) The `200 OK` response to the PRACK request is extended to the CS network.

16) The UPDATE request is generated by the CS network.

19) The `200 OK` response to the UPDATE request is extended to the CS network.

22) The ISUP ACM message is transmitted in the provisional response `180 Ringing` to the INVITE request.

25) The ISUP ANM message is transmitted in the `200 OK` response to the INVITE request.

26) The ACK request is generated by the CS network.

6.1.4. *Session termination*

6.1.4.1. *Session termination initiated by IMS network*

Procedure for the release of the session initiated by the IMS network is displayed in Figure 6.9.

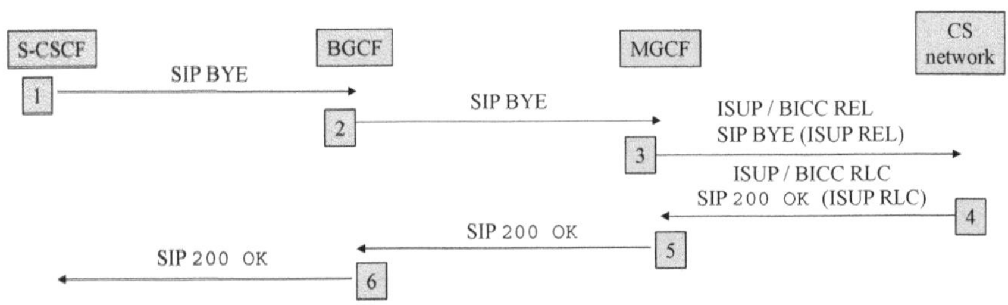

Figure 6.9. *Session clearing initiated by IMS network ISUP, BICC or SIP-I signaling*

1), 2) The UA entity terminates the communication by generating the BYE request.

3) Upon receipt of the BYE request, the MGCF entity removes the end-points of the IMS-MGW gateway and generates the message ISUP/BICC REL to the CS network, in the case of ISUP/BICC signaling.

In the case of SIP-I signaling, the ISUP REL message is carried in the BYE request.

4) In the case of ISUP/BICC signaling, the CS network responds with the message ISUP / BICC RLC.

In the case of SIP-I signaling, the ISUP RLC message is conveyed in the `200 OK` response to the BYE request generated by the MGCF entity.

5), 6) The `200 OK` message is the response to the BYE request generated by the UA entity.

6.1.4.2. *Session termination initiated by CS network*

The procedure for the release of the session initiated by the CS network is described in Figure 6.10.

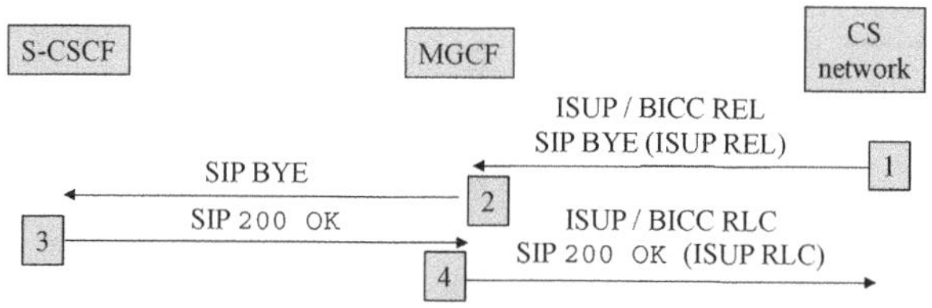

Figure 6.10. *Session clearing initiated by CS network ISUP, BICC or SIP-I signaling*

1) In the case of ISUP/BICC signaling, the MGCF entity receives the message ISUP / BICC REL from the CS network.

In case of SIP-I signaling, the ISUP REL message is carried in the BYE request.

2) The MGCF entity, in turn, generates a BYE request sent to Bob's UA entity and deletes the end-points of the IMS-MGW gateway.

3) The `200 OK` message is the response to the BYE request generated by the MGCF entity.

4) On receiving this message, the MGCF entity generates the message ISUP/ BICC RLC, addressed to the CS network.

In the case of SIP-I signaling, the ISUP RLC message is conveyed in the `200 OK` response to the BYE request generated by the MGCF entity.

6.2. Interconnection with IMS network

6.2.1. *Functional architecture*

The functional architecture of the IMS network implementing the interconnection with other IMS networks is described in Figure 6.11.

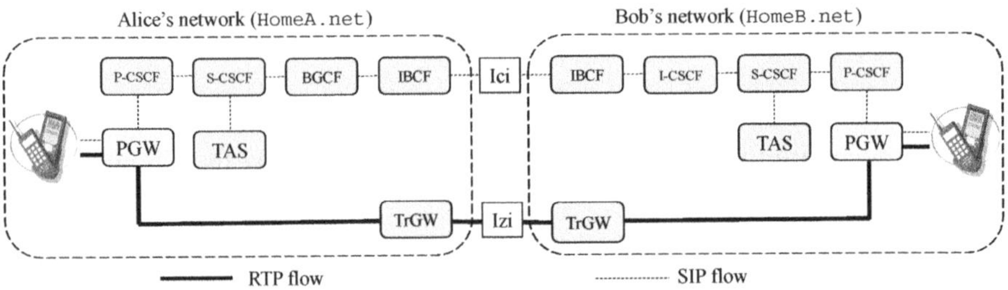

Figure 6.11. *Interconnection with IMS network functional architecture of IMS network*

In case of an outgoing call, the S-CSCF entity detects that the called subscriber belongs to a different domain and forwards the INVITE request to the BGCF entity that retains its search function of the entity in charge of control interconnection.

The interconnection border control function (IBCF) is the gateway that allows the access of the SIP flow to another IMS network.

The IBCF entity can make the translation of IP addresses and port numbers, corresponding to the network address and port translation (NAPT).

The IBCF entity can perform the translation of IP addresses, port numbers and conversion of IPv4 to IPv6, corresponding to the NAPT-PT (Protocol Translation) function.

The IBCF entity performs the masking of the topology of the IMS network,

The IBCF entity performs the withdrawal of some headers of the SIP message based on the rules established by the operator, corresponding to the function topology hiding interconnect gateway (THIG).

The transition gateway (TrGW) is the entity that anchors the RTP stream and allows access traffic to another IMS network.

The TrGW entity may perform filtering, transcoding and NAPT or NAPT-PT translation of the RTP streams under the control of the IBCF entity.

The interconnection between IMS networks is provided by:

– the Ici interface between IBCF entities for the SIP flow;

– the Izi interface between TrGW entities for the RTP stream.

6.2.2. *Session establishment*

Table 6.1 summarizes the IP addresses and port numbers for the different RTP streams.

	Alice's UA	TrGW (`HomeA.net`)
IP address	192.0.2.1	192.0.2.2
Port number	49170	56743
	TrGW (`HomeA.net`)	TrGW (`HomeB.net`)
IP address	178.15.1.1	178.15.1.2
Port number	62111	33248
	TrGW (`HomeB.net`)	Bob's UA
IP address	190.1.15.1	190.1.15.2
Port number	12538	24152

Table 6.1. *RTP flow characteristics*

The procedure for establishing the session is illustrated in Figure 6.12. To simplify the presentation, responses `100 Trying`, `180 Ringing` and precondition mechanisms are avoided.

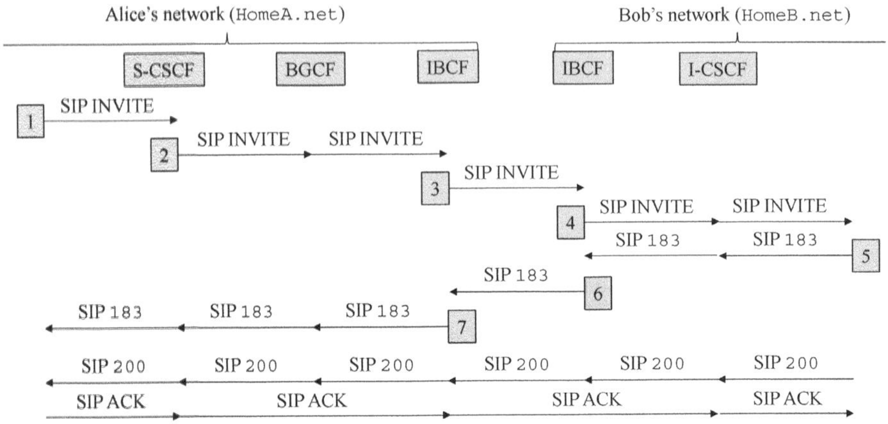

Figure 6.12. *Interconnection with IMS network session establishment*

1) The S-CSCF entity receives from the P-CSCF entity the INVITE message whose associated SDP message contains the characteristics of the RTP streams of Alice's UA entity.

```
INVITE SIP: tel:+4687197378; SIP/2.0
...
Route: <sip:scscf1.operatorHA.net;lr>
...
Content-Type: application/sdp
Content-Length: (...)
...
c=IN IP4 192.0.2.1
m=audio 49170 RTP/AVP 96 97
...
```

2) The S-CSCF entity detects that Alice's and Bob's UA entities are in different domains and transfers the INVITE message to the BGCF entity.

The S-CSCF entity removes its identity from the `Route` header and performs the ENUM resolution on the URI identity of the destination to which it adds the `iotl` parameter indicating the direction of the request (`homeA-Homeb`).

The BGCF entity selects the IBCF entity and forwards the INVITE message without changing the SDP message.

```
INVITE
<sip:+46107197378@operatorHb.net;user=phone;iotl=homeA-
homeB> SIP/2.0
...
Content-Type: application/sdp
Content-Length: (...)
...
c=IN IP4 192.0.2.1
m=audio 49170 RTP/AVP 96 97
...
```

3) The IBCF entity of the `HomeA.net` network selects the IBCF entity of the `HomeB.net` network and forwards the INVITE message including the SDP message which replaces the characteristics of the RTP streams received from Alice's UA entity with those communicated the TrGW entity of the `HomeA.net` network.

```
INVITE
<sip:+46107197378@operatorHb.net;user=phone;iotl=homeA-
homeB> SIP/2.0
...
Content-Type: application/sdp
Content-Length: (...)
...
c=IN IP4 178.15.1.1
m=audio 62111 RTP/AVP 96 97
...
```

4) The IBCF entity of the `HomeB.net` network transfers to the I-CSCF entity the INVITE message including the SDP message which replaces the characteristics of the RTP streams received from the IBCF entity of the `HomeA.net` network with those communicated by the TrGW entity of the `HomeB.net` network.

The I-CSCF entity retrieves from the HSS entity the identity of the S-CSCF entity and forwards the INVITE message without changing the SDP message.

```
INVITE
<sip:+46107197378@operatorHb.net;user=phone;iotl=homeA-
homeB> SIP/2.0
...
Content-Type: application/sdp
Content-Length: (...)
...
c=IN IP4 190.1.15.2
m=audio 12538 RTP/AVP 97 98
...
```

5) I-CSCF entity receives the provisional response `183 Session in Progress` from the S-CSCF entity with the associated SDP message containing the RTP streams characteristics of Bob's UA entity.

```
SIP/2.0 183 Session Progres
...
Content-Type: application/sdp
Content-Length: (...)
...
c=IN IP4 190.1.15.1
m=audio 24152 RTP/AVP 97 98
...
```

The I-CSCF entity forwards to the IBCF entity of the `HomeB.net` network the provisional response `183 Session in Progress` without changing the SDP message.

6) The IBCF entity of the `HomeB.net` network transfers to the IBCF entity of the `HomeA.net` network the provisional response `183 Session in Progress` including the SDP message which replaces the characteristics of the RTP streams received from Bob's UA entity with those provided by the TrGW entity of the `HomeB.net` network.

```
SIP/2.0 183 Session Progress
...
Content-Type: application/sdp
Content-Length: (...)
...
c=IN IP4 178.15.1.2
m=audio 33248 RTP/AVP 97 98
...
```

7) The IBCF entity of the `HomeA.net` network transfers to the BGCF entity the provisional response `183 Session in Progress` including

the SDP message which replaces the characteristics of RTP streams received from the IBCF entity of the `HomeB.net` network with those communicated by the TrGW entity of the `HomeA.net` network.

```
SIP/2.0 183 Session Progress
...
Content-Type: application/sdp
Content-Length: (...)
...
c=IN IP4 192.0.2.2
m=audio 56743 RTP/AVP 97 98
...
```

7

Handover

7.1. Introduction

Handover is the mechanism that maintains the current session (the voice or the conversational video communication) when the mobile changes the radio cell.

During the inter-system handover, cell change takes place without mobile network changes.

During the intra-system handover, serving gateway (SG) and PDN gateway (PGW) entities play the role of an anchor point, which will hide the terminal mobility from the IMS network.

There are two types of intra-system handover:

– X2-based handover, from the name of the interface between the eNB entities;

– S1-based handover, from the name of the interface between the MME (Mobility Management Entity) and eNB entities.

X2-based handover occurs if both eNB entities, the source and the target, belong to the same group (pool) and the X2-AP interface has been enabled.

During X2-based handover, the MME entity does not change, which is not the case of SGW entity that can be relocated or not.

S1-based handover occurs for the following four cases:

– X2-AP interface between two eNB entities of the same group was not activated, the MME must not be relocated and the SGW entity may be relocated (scenario described in section 7.3.1);

– both eNB entities belong to two different groups, the MME entity must be relocated and the SGW entity may be relocated (scenario described in section 7.3.2) or not.

During the inter-system handover, cell change is accompanied by a change of network, the mobile being transferred from 4G network to 2G or 3G networks.

There are two types of inter-system handover:

– packet-switched (PS-PS) handover, for which the mobile does not change the mode when changing the network;

– circuit-switched (PS-CS) handover, for which the mobile is also changing the mode.

In the case of a voice or conversational video communication, PS-PS inter-system handover occurs only if the voice over high speed packet access (VoHSPA) service is available on 3G network.

During PS-PS inter-system handover, the SGW entity acts as an anchor point, which will hide the terminal mobility to the IMS network.

PS-CS inter-system handover is described relative to the service centralization and continuity in Chapter 8.

The handover procedure is activated by the source eNB entity from measurements on the radio signal made by:

– the eNB entity for the uplink direction;

– the mobile for the downlink direction, for the serving and neighboring cells. These measures are recovered by the eNB entity in RRC *MeasurementReport* messages.

The handover procedure takes place in three phases:

– the preparation phase corresponding to the decision of cell change and resource reservation;

– the execution phase corresponding to the mobile connection to the target eNB entity;

– the completion phase corresponding to the establishment of final bearers and the release of the old resources.

The intra-system handover is carried out with a disconnection of the mobile with the eNB entity, whose duration depends on the following:

– the synchronization of the mobile on the primary synchronization signal (PSS) and secondary synchronization signal (SSS) transmitted by the target eNB entity;

– the discovery of the physical random access channel (PRACH);

– the random access procedure;

– the connection procedure to the target eNB entity.

During the preparation of the intra-system handover, a unidirectional and temporary bearer is created between the source and target eNB entities for the downward direction.

During the execution of the handover, the incoming data are forwarded to the target eNB entity that stores and delivers it to the mobile when it connects.

For voice and conversational video communication, this operation causes an increase in packet jitter, which is corrected using the real-time transport protocol (RTP), to a certain limit. If the jitter of packets from end to end becomes too large, the packets are dropped.

7.2. Handover based on X2

7.2.1. *Handover based on X2 without relocation*

The functional architecture of the handover based on X2 without relocation of the SGW entity is described in Figure 7.1.

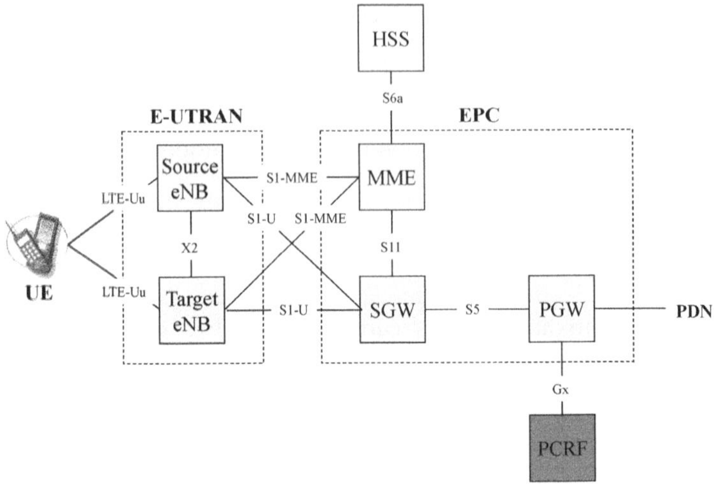

Figure 7.1. *Handover based on X2 without relocation functional architecture*

The X2-based handover procedure without relocation of the SGW entity is described in Figure 7.2:

– the preparation phase includes messages 1 and 2;

– the implementation phase includes messages 3 to 5;

– the completion phase includes messages 6 to 15.

1) On receipt of the RRC *MeasurementReport* message, the source eNB entity decides to perform a cell change and transmits to the target eNB entity the message X2-AP HANDOVER REQUEST containing the context of the mobile, particularly the tunnel endpoint identifier (TEID) of the S1 bearer provided by the SGW entity.

2) The target eNB entity responds with the message X2-AP HANDOVER REQUEST ACK, containing the technical characteristics of the radio interface in the information element. Handover command and the TEID identifier of the unidirectional X2 temporary bearer built between source and target eNB entities.

3) The source eNB entity eNB starts the execution phase by sending to the mobile the message RRC *ConnectionReconfiguration* containing information element handover.

4) The source eNB entity sends to the target eNB the sequence numbers of the packet data convergence protocol (PDCP) in the message X2-AP SN STATUS TRANSFER and incoming data that have not been acknowledged by the mobile and that the target eNB entity will store until the mobile is able to receive them.

5) When the RRC connection is established between the mobile and the target eNB entity, the mobile transmits the message RRC *ConnectionReconfigurationComplete*, which will trigger the transfer of incoming data stored by the target eNB entity to the mobile.

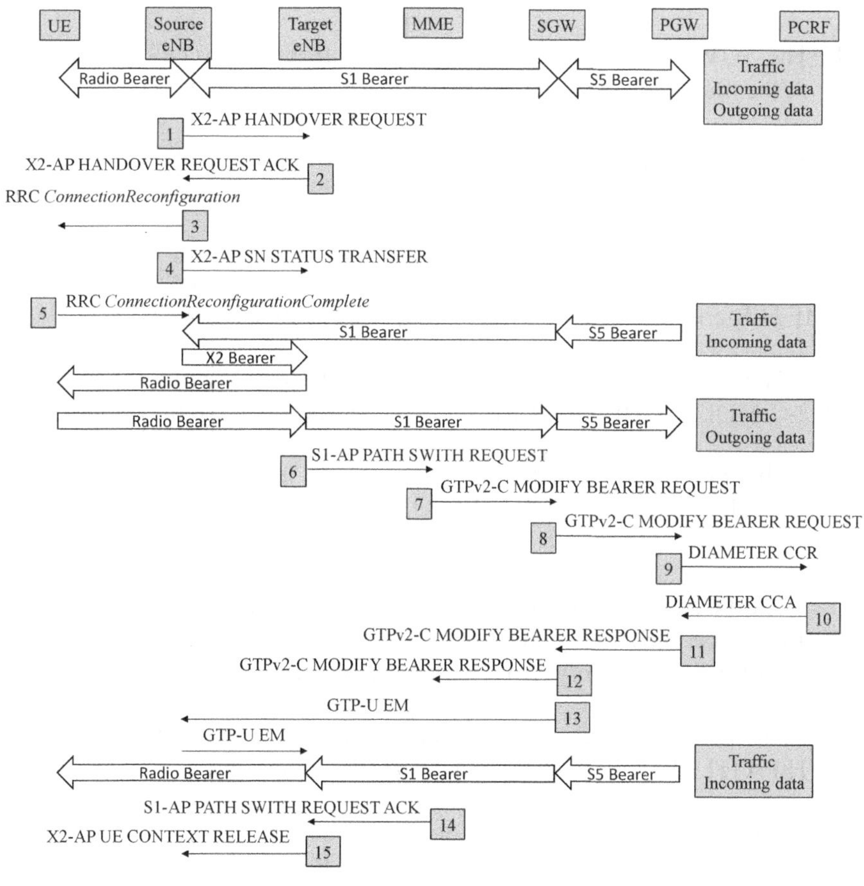

Figure 7.2. *Handover based on X2 without relocation procedure*

6) The target eNB entity starts the completion phase and transmits to the MME entity the message S1-AP SWITH PATH REQUEST containing the following information:

– the TEID identifier of the S1 bearer that the SGW entity will use in the GTP-U header when it sends traffic to the target eNB entity;

– the E-UTRAN cell global identifier (ECGI).

7) The MME determines that the SGW entity shall not change and transfers this information to the SGW entity in the message GTPv2-C MODIFY BEARER REQUEST.

8) The SGW entity transfers the ECGI identifier of the cell to the PGW entity in the message GTPv2-C MODIFY BEARER REQUEST.

9) The PGW entity transfers the ECGI identifier of the cell to the policy and charging rules function (PCRF) in the message DIAMETER credit-control-request (CCR).

10) The PCRF entity responds to the PGW entity with the message DIAMETER credit-control-answer (CCA) to acknowledge the request.

11) The PGW entity responds to the SGW entity with the message GTPv2-C MODIFY BEARER RESPONSE to acknowledge the request.

12) The SGW entity responds to the MME entity with the message GTPv2-C MODIFY BEARER RESPONSE message to acknowledge the request.

13) The target eNB entity transmits to the mobile the traffic received from the source eNB entity until it receives the GTP-U EM (End Marker) messages of the SGW entity, from which the traffic will be received directly.

14) The MME responds to the target eNB entity with the message S1-AP SWITH PATH REQUEST ACK to acknowledge the request.

15) The target eNB entity informs the source eNB entity that the handover was achieved with the message X2-AP EU CONTEXT RELEASE so that it releases the context associated with the mobile.

7.2.2. Handover based on X2 with relocation

The functional architecture of the handover based on X2 with relocation of the SGW entity is shown in Figure 7.3.

The X2-based handover procedure with relocation of SGW entity for the completion phase is illustrated in Figure 7.4.

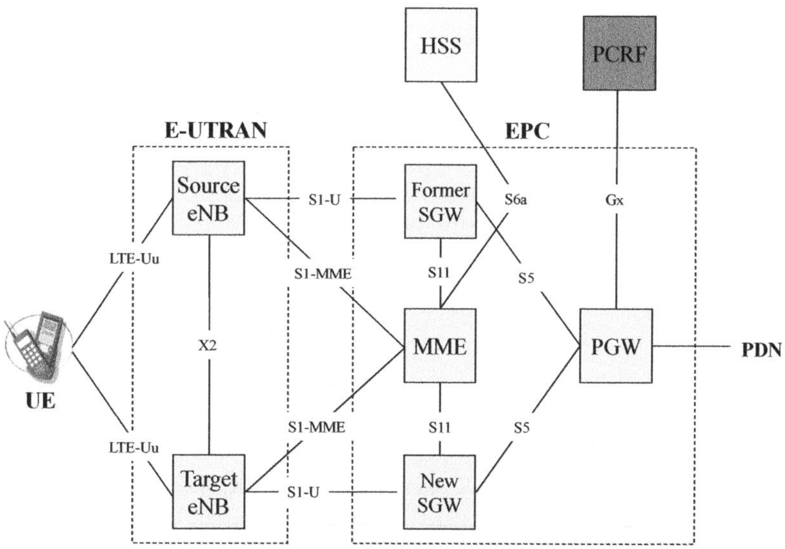

Figure 7.3. *Handover based on X2 with relocation functional architecture*

The preparation and execution phases are identical to those described for the handover based on X2 without relocation.

1) The target eNB entity starts the completion phase and transmits to the MME entity the message S1-AP SWITH PATH REQUEST containing TEID and ECGI identifiers.

2) The MME entity determines that the SGW entity must be relocated and transfers this information as well as the IP address of the entity PGW to the new SGW entity in the message GTPv2-C CREATE SESSION REQUEST.

3) The new SGW entity forwards the ECGI identifier of the cell to the PGW entity in the message GTPv2-C MODIFY BEARER REQUEST that contains the TEID identifier that the PGW entity will use when it sends traffic to the new SGW entity.

4) The PGW entity forwards the ECGI identifier of the cell to the PCRF entity in the DIAMETER CCR message.

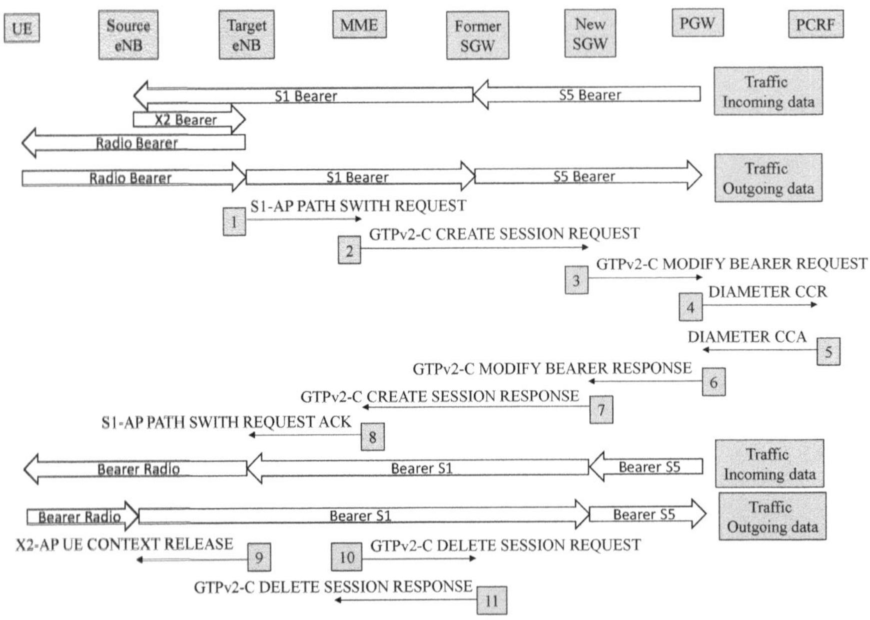

Figure 7.4. *Handover based on X2 with relocation completion phase*

5) The PCRF entity responds to the PGW entity with the DIAMETER CCA message to acknowledge the request.

6) The PGW entity responds to the SGW entity with the message GTPv2-C MODIFY BEARER RESPONSE containing the TEID identifier that the new SGW entity will use when it sends traffic to the PGW entity.

7) The new SGW entity responds to the MME entity with the message GTPv2-C CREATE SESSION RESPONSE containing the TEID identifier

that the target eNB entity will use in the GTP-U header when sending traffic to the new SGW entity.

8) The MME entity responds to the target eNB entity with the message S1-AP SWITH PATH REQUEST ACK containing the TEID identifier provided by the new SGW entity.

9) The target eNB entity informs the source eNB entity that the handover was achieved with the message X2-AP EU CONTEXT RELEASE that it releases the context associated with the mobile.

10) The MME entity controls the release of the context of the former SGW entity with the message GTPv2-C DELETE SESSION REQUEST.

11) The former SGW entity responds to the MME entity with the message GTPv2-C DELETE SESSION REQUEST to acknowledge the request.

7.3. Handover based on S1

7.3.1. *Handover based on S1 without relocation*

The functional architecture for the handover based on S1 without relocation of the MME and SGW entities corresponds to Figure 7.1, for which the X2 interface between the source and target eNB entities is deactivated.

The MME is no longer transparent to the handover mechanism and acts as a signaling relay for the handover control between the source and target eNB entities.

The temporary bearer built between the source and target eNB entities for incoming data passes through the SGW entity.

The handover procedure based on S1 without relocation of the MME and SGW entities is given in Figure 7.5:

– the preparation phase includes messages 1 to 6;

– the execution phase includes messages 7 to 10;

– the completion phase includes messages 11 to 22.

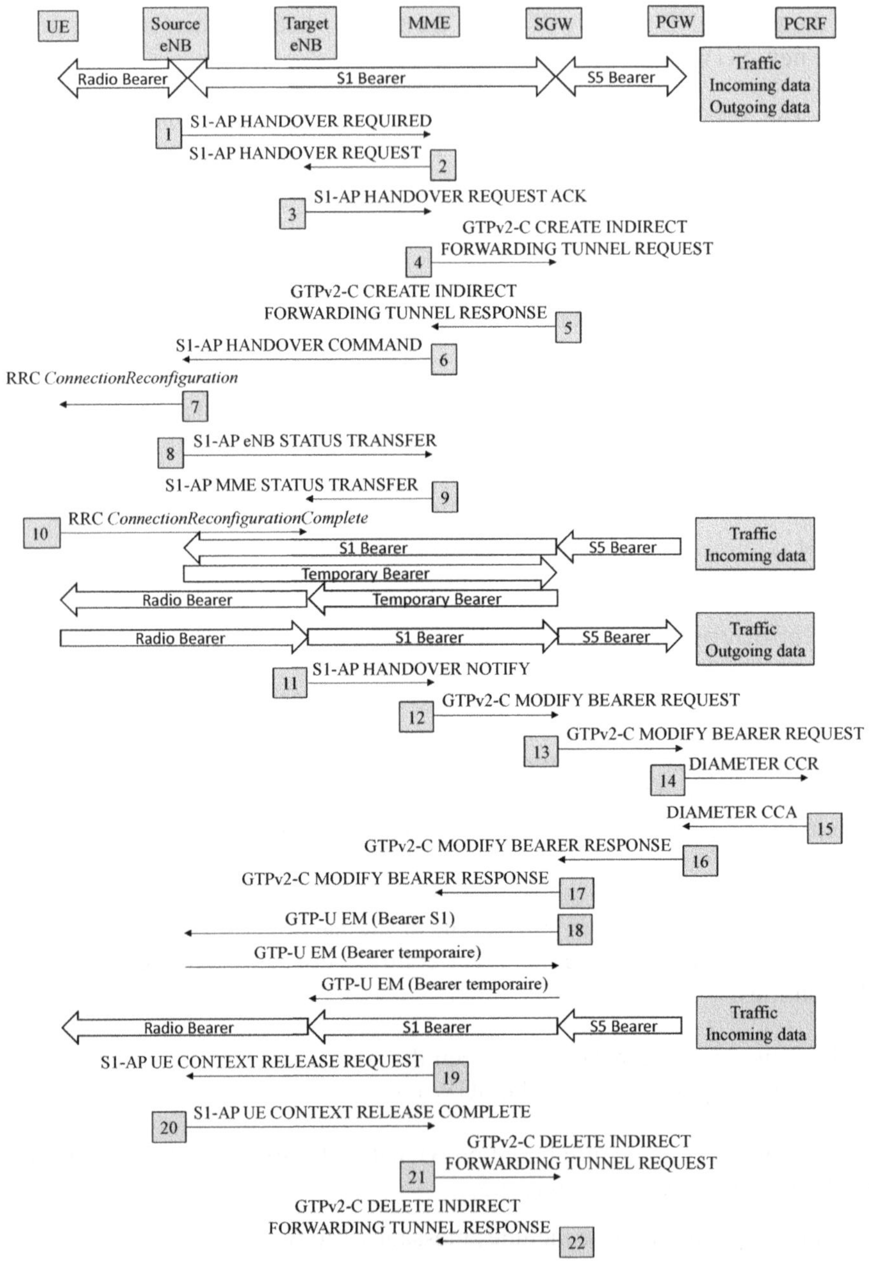

Figure 7.5. *Handover based on S1 without relocation procedure*

1) The source eNB entity initiates the preparation phase of the handover by sending the message S1-AP HANDOVER REQUIRED to the MME entity.

2) The MME entity sends to the target eNB entity the message S1-AP HANDOVER REQUEST to perform the reservation of resources.

3) The target eNB entity responds to the MME entity with the message S1-AP HANDOVER REQUEST ACK containing the following information:

– the TEID identifier of the temporary bearer that the SGW entity will use in the GTP-U header when it sends traffic to the target eNB entity;

– the TEID identifier of the S1 bearer that the SGW entity will use in GTP-U header when it sends traffic to the target eNB entity;

– the technical characteristics of the radio interface in the information element handover command.

4) The MME entity forwards the TEID identifier relating to the temporary bearer to the SGW entity in the message GTPv2-C CREATE INDIRECT FORWARDING TUNNEL REQUEST.

5) The SGW entity acknowledges the creation of the temporary bearer with the message GTPv2-C CREATE INDIRECT FORWARDING TUNNEL RESPONSE containing the TEID identifier of the temporary bearer that the source eNB entity will use in the GTP-U header when it will send traffic to the SGW entity.

6) The preparation phase ends when the MME entity sends to the source eNB entity the message S1-AP HANDOVER COMMAND containing the following information:

– the TEID identifier of the temporary bearer that the source eNB entity will use in the GTP-U header when it sends traffic to the SGW entity;

– the technical characteristics of the radio interface of the target eNB entity in the information element handover command.

7) The source eNB entity starts the execution phase by sending to the mobile the RRC *ConnectionReconfiguration* message containing the information element *Handover Command.*

8) The source eNB entity transmits to the MME entity the sequence numbers of the PDCP protocol with the message SI-AP SN STATUS TRANSFER.

9) The MME entity forwards the sequence numbers to the target eNB entity in the message SI-AP MME STATUS TRANSFER.

Incoming data that have not been acknowledged by the mobile which the target eNB entity will store until the mobile is able to receive are transmitted in the temporary bearer.

10) When the RRC connection is established between the mobile and the target eNB entity, the mobile transmits the message RRC *ConnectionReconfigurationComplete*, which will trigger the transfer of incoming data stored by the target eNB entity to the mobile.

11) The completion phase starts when the target eNB entity notifies the MME entity with the message S1-AP HANDOVER NOTIFY containing the ECGI identity of the cell.

12) The MME entity forwards to the SGW entity the message GTPv2-C MODIFY BEARER REQUEST containing the ECGI identity of the cell and the TEID identifier of the S1 bearer that the SGW entity will use in the GTP-U header when it sends traffic to the target eNB entity.

13) The SGW entity transfers the ECGI identifier of the cell to the PGW entity in the message GTPv2-C MODIFY BEARER REQUEST.

14) The PGW entity transfers the ECGI identifier of the cell to the PCRF entity in the DIAMETER CCR message.

15) The PCRF entity responds to the PGW entity with the DIAMETER CCA message to acknowledge the request.

16) The PGW entity responds to the SGW entity with the message GTPv2-C MODIFY BEARER RESPONSE to acknowledge the request.

17) The SGW entity responds to the MME entity with the message GTPv2-C MODIFY BEARER RESPONSE to acknowledge the request.

18) The target eNB entity transmits to the mobile the traffic received from the source eNB entity by the temporary bearer until it receives a

GTP-U EM message from the SGW entity from which the traffic will be received directly from the S1 bearer.

19) The release of the context of the mobile at the source eNB entity is triggered by the MME entity by sending the message S1-AP UE CONTEXT RELEASE REQUEST.

20) The source eNB entity responds to the MME entity with the message S1-AP and acknowledges the message received.

21) The release of the temporary bearer at the SGW entity is triggered by the MME entity by sending the message GTPv2-C DELETE INDIRECT FORWARDING TUNNEL REQUEST.

22) The SGW entity responds to the MME entity with the message GTPv2-C DELETE INDIRECT FORWARDING TUNNEL RESPONSE to acknowledge the received message.

7.3.2. *Handover based on S1 with relocation*

The functional architecture for the handover based on S1 with the relocation of the MME and SGW entities is described in Figure 7.6.

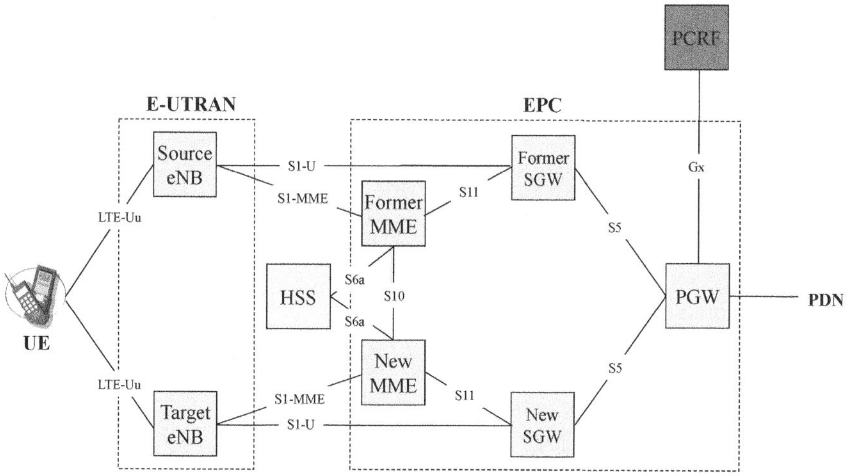

Figure 7.6. *Handover based on S1 with relocation functional architecture*

7.3.2.1. *Preparation phase*

The S1-based handover procedure with the relocation of the MME and SGW entities is described in Figure 7.7 for the preparation phase.

1) The source eNB entity initializes the preparation phase of the handover by sending the message S1-AP HANDOVER REQUIRED to the former MME entity.

2) The former MME entity selects a new MME entity and transmits the message GTPv2-C FORWARD RELOCATION REQUEST containing the following information:

– the IP address of the PGW entity;

– the TEID identifier relative to the S5 bearer that the new SGW entity will use in the GTP-U header when it sends traffic to the PGW entity.

3) The new MME decides to relocate the SGW entity and passes to the new SGW entity the message GTPv2-C CREATE SESSION REQUEST containing the information received.

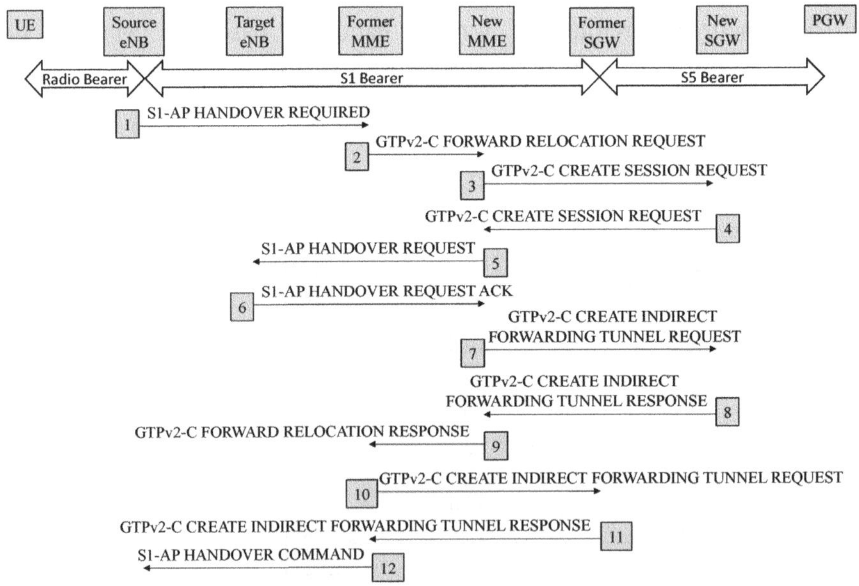

Figure 7.7. *Handover based on S1 with relocation preparation phase*

4) The new SGW entity responds with the message GTPv2-C CREATE SESSION REQUEST containing the TEID identifier of the S1bearer that the target eNB entity will use in the GTP-U header when it sends traffic to the new SGW entity.

5) The new MME entity transmits to the target eNB entity the message S1-AP HANDOVER REQUEST to perform the reservation of resources.

6) The target eNB entity responds to the new MME entity with the message S1-AP HANDOVER REQUEST ACK containing the following information:

– the TEID identifier of the temporary bearer that the new SGW entity will use in the GTP-U header when it sends traffic to the target eNB entity;

– the TEID identifier of the S1 bearer that the new SGW entity will use in the GTP-U header when it sends traffic to the target eNB entity;

– the technical characteristics of the radio interface in the information element handover command.

7) The new MME entity forwards the TEID identifier relating to the temporary bearer to the SGW entity in the message GTPv2-C CREATE INDIRECT FORWARDING TUNNEL REQUEST.

8) The new SGW entity acknowledges the creation of the temporary bearer by the message GTPv2-C CREATE INDIRECT FORWARDING TUNNEL RESPONSE containing the TEID identifier of the temporary bearer that the former SGW entity will use in the GTP-U header when it will send traffic to the new SGW entity.

9) The new MME entity responds to the former MME entity with the message GTPv2-C FORWARD RELOCATION REQUEST containing the following information:

– the TEID identifier of the temporary bearer that the former SGW entity will use in the GTP-U header when it sends traffic to the new SGW entity;

– the technical characteristics of the radio interface of the target eNB entity in the information element Handover Command.

10) The former MME entity forwards the TEID identifier relating to the temporary bearer to the former SGW entity in the message GTPv2-C CREATE INDIRECT FORWARDING TUNNEL REQUEST.

11) The former SGW entity acknowledges the creation of the temporary bearer with the message GTPv2-C CREATE INDIRECT FORWARDING TUNNEL RESPONSE containing the TEID identifier of the temporary bearer that the source eNB entity will use in the GTP-U header when it will send traffic to the former entity SGW.

12) The preparation phase ends when the former MME entity transmits to the source eNB entity the message S1-AP HANDOVER COMMAND containing the following information:

– the TEID identifier from the temporary bearer that the source eNB entity will use in the GTP-U header when it sends traffic to the former SGW entity;

– the technical characteristics of the radio interface of the target eNB entity in the information element handover command.

7.3.2.2. *Execution phase*

The S1-based handover procedure with relocation of the MME and SGW entities is described in Figure 7.8 for the execution phase.

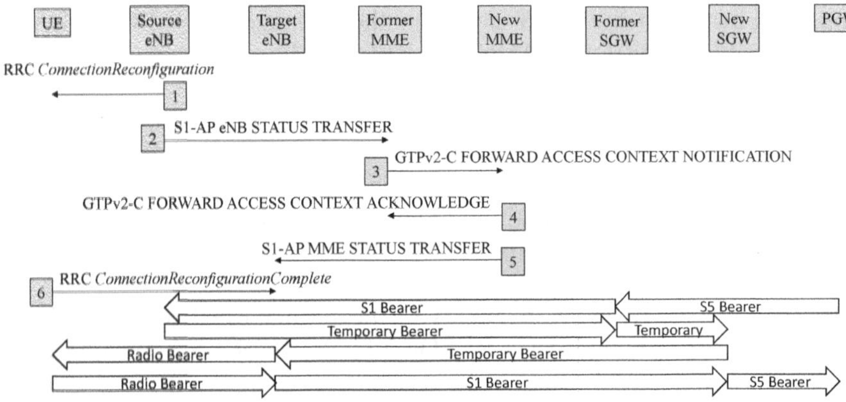

Figure 7.8. *Handover based on S1 with relocation execution phase*

1) The source eNB entity starts the execution phase by sending to the mobile the RRC *ConnectionReconfiguration* message containing the information element handover command.

2) The source eNB entity transmits to the former MME entity the sequence numbers of the PDCP protocol with the message SI-AP SN STATUS TRANSFER.

3) The former MME entity forwards the sequence numbers to the new MME entity in the message GTPv2-C FORWARD ACCESS CONTEXT NOTIFICATION.

4) The new MME entity responds to the former MME entity with the message GTPv2-C FORWARD ACCESS CONTEXT ACKNOWLEDGE to acknowledge the message received.

5) The new MME entity forwards the sequence numbers to the target eNB entity in the message SI-AP MME STATUS TRANSFER.

Incoming data that have not been acknowledged by the mobile which the target eNB entity will store until the mobile is able to receive are transmitted in the temporary bearer.

6) When the RRC connection is established between the mobile and the target eNB entity, the mobile transmits the RRC *ConnectionReconfigurationComplete* message, which will trigger the transfer of incoming data stored by the target eNB entity to the mobile.

7.3.2.3. *Completion phase*

The S1-based handover procedure with the relocation of the MME and SGW entities is shown in Figure 7.9 for the completion phase.

1) The completion phase starts when the target eNB entity notifies the new MME entity with the message S1-AP HANDOVER NOTIFY containing the ECGI identity of the cell.

2) The new MME entity informs the former MME entity that the handover is completed with the message GTPv2-C FORWARD RELOCATION COMPLETE NOTIFICATION.

3) Former MME entity responds to the new MME entity with the message GTPv2-C FORWARD RELOCATION COMPLETE ACKNOWLEDGE to acknowledge the message received.

4) The new MME entity forwards to the new SGW entity in the message GTPv2-C MODIFY BEARER REQUEST containing the ECGI identity of the cell and the TEID identifier of the S1 bearer that the new SGW entity will use in the GTP-U header when it sends traffic to the target eNB entity.

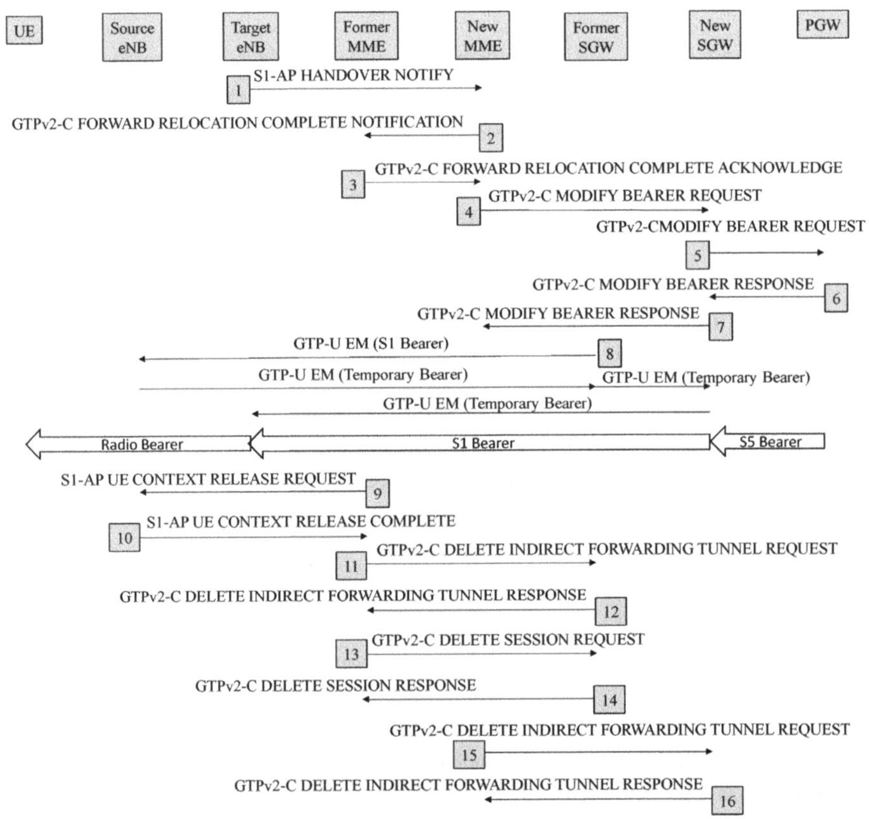

Figure 7.9. *Handover based on S1 with relocation completion phase*

5) The new SGW entity forwards the ECGI identifier of the cell to the PGW entity in the message GTPv2-C MODIFY BEARER REQUEST.

The PGW entity transfers the ECGI identifier of the cell to the PCRF entity in the DIAMETER RRC message.

The PCRF entity responds to the PGW entity with the DIAMETER CCA message to acknowledge the request.

6) The PGW entity responds to the new SGW entity with the message GTPv2-C MODIFY BEARER RESPONSE to acknowledge the request.

7) The new SGW entity responds to the new MME entity with the message GTPv2-C MODIFY BEARER RESPONSE to acknowledge the request.

8) The target eNB entity transmits to the mobile the traffic received from the source eNB entity by the temporary bearer until it receives the message GTP-U EM of the new SGW entity, from which the traffic will be received directly by the S1 bearer.

9) The release of the context of the mobile at the source eNB entity is triggered by the former MME entity by sending the message S1-AP EU CONTEXT RELEASE REQUEST.

10) The source eNB entity responds to the former MME entity with the message S1-AP EU CONTEXT RELEASE COMPLETE to acknowledge the received message.

11) The release of the temporary bearer at the former SGW entity is triggered by the former MME entity by sending the message GTPv2-C DELETE INDIRECT FORWARDING TUNNEL REQUEST.

12) The former SGW entity responds to the former MME entity with the message GTPv2-C DELETE INDIRECT FORWARDING TUNNEL RESPONSE to acknowledge the received message.

13) The release of the context at the former SGW entity is triggered by the former MME entity by sending the message GTP-C DELETE SESSION REQUEST.

14) The former SGW entity responds to the former MME entity with the message GTP-C DELETE SESSION RESPONSE to acknowledge the received message.

15) The release of the temporary bearer at the new SGW entity is triggered by the new MME entity by sending the message GTPv2-C DELETE INDIRECT FORWARDING TUNNEL REQUEST.

16) The new SGW entity responds to the new MME entity with the message GTPv2-C DELETE INDIRECT FORWARDING TUNNEL RESPONSE to acknowledge the received message.

7.4. PS-PS inter-system handover

7.4.1. *Functional architecture*

PS-PS inter-system handover impacts the serving GPRS support node (SGSN) of 2G/3G mobile network and possibly the radio network controller (RNC) if the feature Direct Tunnel is implemented.

The SGSN entity plays for 2G/3G mobile networks the same role as the MME entity for a 4G mobile network.

The RNC entity plays for a 3G mobile network the same role as the eNB entity regarding the control of the mobile connection and the allocation of resources on the radio interface.

The feature Direct Tunnel, specific to a 3G mobile network, allows for direct transfer traffic between the RNC and gateway GPRS support node (GGSN) entities without passing through the SGSN entity.

The GGSN entity plays for 2G/3G mobile networks the same role as the PGW entity for 4G mobile network.

The SGW entity ensures the anchor point of 2G/3G mobile networks:

– anchoring the traffic is carried by the SGSN entity that connects to the SGW entity if the feature Direct Tunnel is not available;

– anchoring the traffic is carried by the RNC entity that connects to the SGW entity if the feature Direct Tunnel is available.

The functional architecture for the PS-PS inter-system handover is described in Figure 7.10.

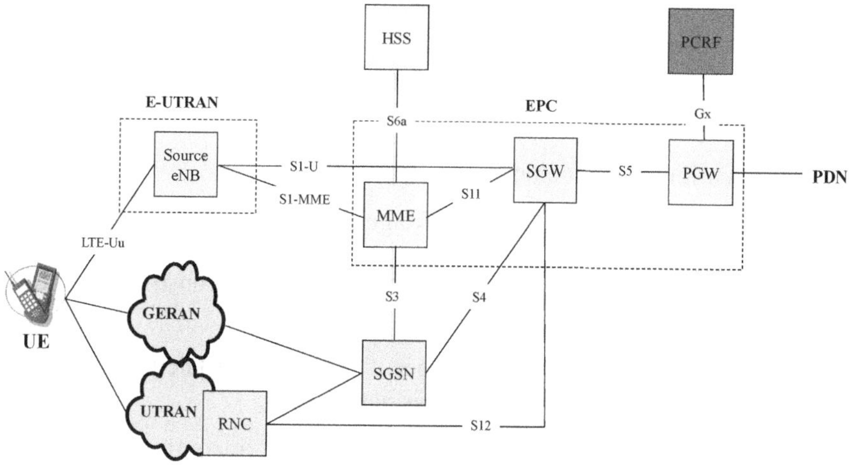

Figure 7.10. *PS-PS inter-system handover functional architecture*

The S3 interface is the reference point between the SGSN and MME entities:

– this interface supports GTPv2-C (GPRS Tunnel Protocol Control) signaling;

– this interface allows the exchange of messages related to the management of PS-PS inter-system handover.

The S4 interface is the reference point between the SGW and SGSN entities:

– this interface is deployed if the feature Direct Tunnel is not available;

– this interface supports GTPv2-C signaling for the control plane and GTP-U (GPRS Tunnel Protocol User) tunneling for the traffic plane;

– GTPv2-C signaling ensures the construction of S4 bearer built between the SGW and SGSN entities.

The S12 interface is the reference point between the SGW and RNC entities:

– this interface is deployed if the feature Direct Tunnel is available;

– this interface supports the GTP-U tunneling for the traffic plane.

7.4.2. Procedure

The procedure for PS-Ps inter-system handover is shown in Figure 7.11:

– the preparation phase includes messages 1 to 8;

– the execution phase includes messages 9 and 10;

– the completion phase includes messages 11 to 21.

During the preparation phase, a temporary bearer is constructed to route the incoming data:

– the temporary bearer may be direct if the data is routed directly from the eNB entity to the 2G/3G radio access network;

– the temporary bearer may be indirect if the incoming data passes through the SGW and SGSN entities.

The procedure for PS-PS inter-system handover is elaborated with the following assumptions in hand:

– the temporary bearer built between the eNB and RNC entities is a direct bearer;

– the feature Direct Tunnel is not available.

1) to 4) The messages are equivalent to messages 1 to 4 described for the preparation phase of the handover based on S1 with relocation of the MME and SGW entities.

5) The SGSN entity transmits to the RNC entity the message RANAP RELOCATION REQUEST to reserve resources in the radio access network.

6) The RNC entity responds with the message RANAP RELOCATION REQUEST ACK to acknowledge the request.

7), 8) The messages are equivalent to messages 11 and 12 described for the preparation phase of the handover based on S1 with relocation of the MME and SGW entities.

9) The message is equivalent to the message 1 described for the execution phase of the handover based on S1 with relocation of the MME and SGW entities.

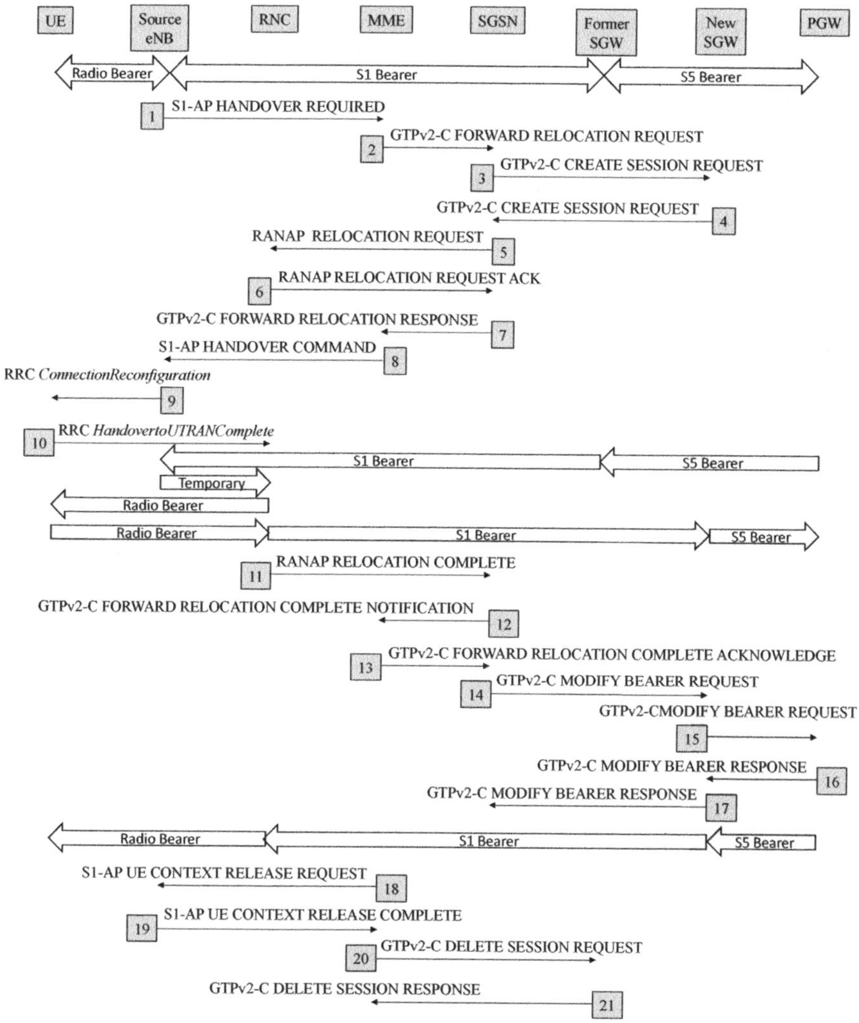

Figure 7.11. *PS-PS inter-system handover procedure*

10) The mobile confirms its connection to the RNC entity with the RRC *HandovertoUTRANComplete* message.

11) The RNC entity informs the SGSN entity about the connection of the mobile with the message RANAP RELOCATION COMPLETE.

12) to 17) The messages are equivalent to messages 2 to 7 described for the completion phase of the handover based on S1 with relocation of the MME and SGW entities.

18), 19) The messages are equivalent to messages 9 and 10 described the completion phase of the handover based on S1 with relocation of the MME and SGW entities.

20), 21) The messages are equivalent to messages 13 and 14 described for the completion phase of the handover based on S1 with relocation of the MME and SGW entities.

8

Roaming

8.1. Functional architecture

8.1.1. *Roaming applied to the EPS network*

The functional architecture of roaming applied to the evolved packet system (EPS) is described in Figure 8.1.

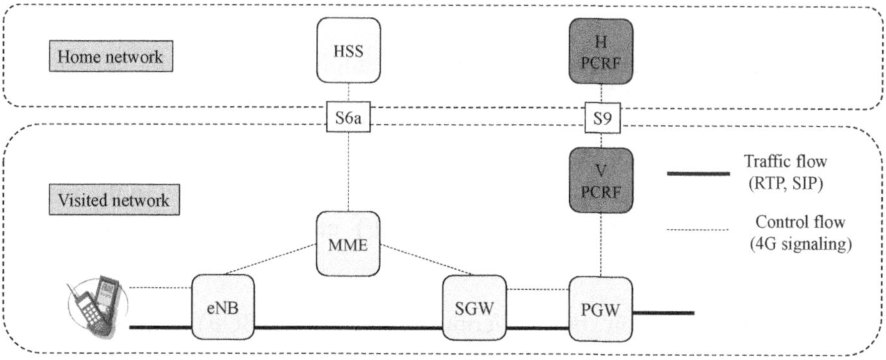

Figure 8.1. *Roaming applied to the EPS network functional architecture*

The functions of the evolved packet core (EPC) and the evolved UMTS terrestrial radio access network (E-UTRAN) are located in the visited network.

The mobility management entity (MME) located in the visited network is connected to the home subscriber server (HSS) located in the home network via the S6a interface.

The S9 interface between the home policy charging and rules function (H-PCRF) and the V-PCRF (Visited PCRF) entity is optional. If the S9 interface is not deployed, the rules applying to mobile traffic in roaming are stored in the subscription profile repository (SPR) associated with the V-PCRF entity.

During the mobile registration, the S9 interface carries the DIAMETER messages of the Gx interface exchanged between V-PCRF and PGW entities for the establishment of the default support (default bearer) assigned to the SIP flow.

During the establishment of the session, S9 interface carries the DIAMETER messages of the Rx interface exchanged between the V-PCRF entity and the proxy call session control function (P-CSCF) in the IMS for the establishment of the dedicated bearer allocated to the RTP stream.

8.1.2. *Roaming applied to the IMS network*

The functional architecture of roaming applied to the IMS network is described in Figure 8.2, when the RTP stream passes through the home network.

The roaming interface is between the P-CSCF entity located in the visited network and the S-CSCF (Serving CSCF) entity located in the home network.

The interconnection border control function (IBCF) can make the translation of IP addresses and port numbers, corresponding to the network address and port translation (NAPT).

The IBCF entity can perform the translation of IP addresses, port numbers and conversion of IPv4 to IPv6, corresponding to the NAPT-PT (Protocol Translation) function.

The IBCF entity performs the withdrawal of some headers of the SIP message based on the rules established by the operator, corresponding to the function topology hiding interconnect gateway (THIG).

The transition gateway (TrGW) may perform filtering, transcoding and NAPT or NAPT-PT translation of the RTP streams under the control of the IBCF entity.

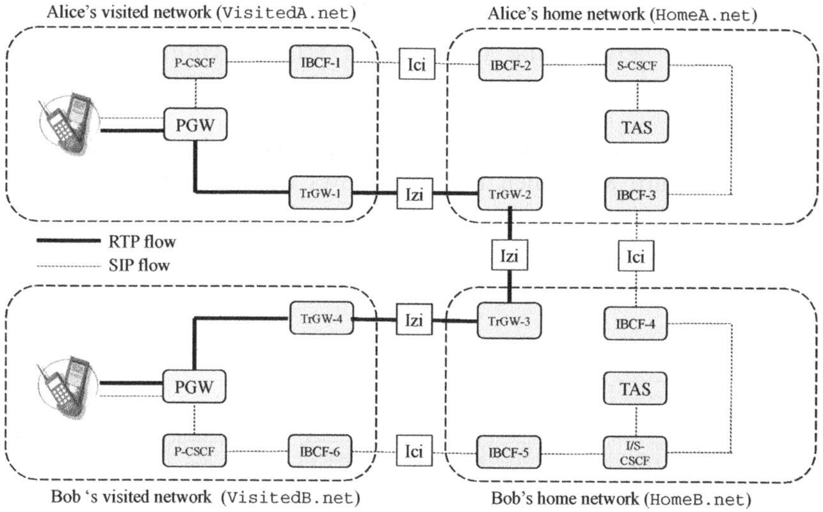

Figure 8.2. *Roaming applied to the IMS network: nominal routeing*

The IBCF entity of Alice's home network (Bob's home network, respectively) consists of two IBCF-2 and-3 IBCF instances (IBCF-4 and IBCF-5, respectively).

The IBCF entity of Alice's home network (Bob's home network, respectively) controls the TrGW-2 entity (TrGW-3, respectively).

Roaming interfaces between the home network and the visited network are provided by:

– Ici interfaces required for SIP flow between IBCF-1 and IBCF-2 instances for Alice's network and between IBCF-5 and IBCF-6 instances for Bob's network;

– Izi interfaces for the RTP stream between TrGW-1 and TrGW-2 entities for Alice's network and between TrGW-3 and TrGW-4 entities for Bob's network.

The interfaces for the interconnection between IMS networks performing, on the one hand, the outgoing call and, on the other hand, the incoming call is provided by:

– Ici interface between IBCF-3 and IBCF-4 instances for the SIP flow;

– Izi interface between the TrGW-2 and TrGW-3 entities for the RTP stream.

The functional architecture of roaming applied to the IMS network is described in Figure 8.3, when the RTP flows does not pass through the home network of the caller, corresponding to the optimal media routeing (OMR).

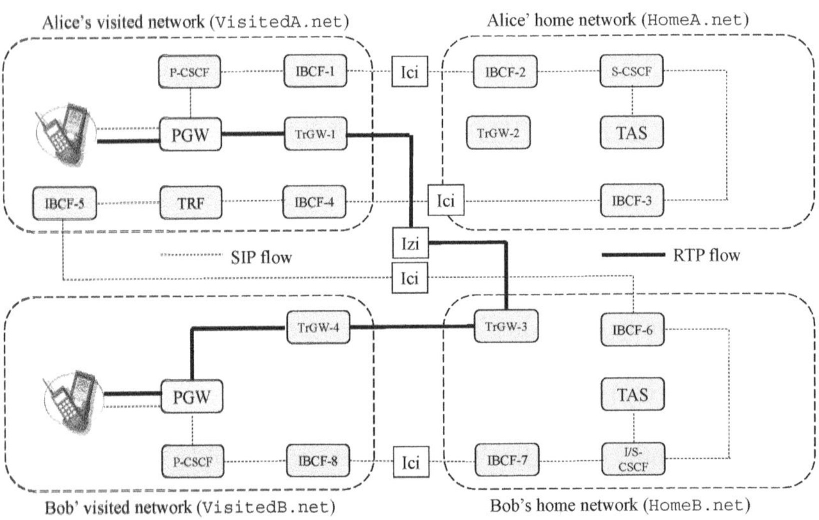

Figure 8.3. *Roaming applied to IMS network: optimal routeing*

The IBCF entity of Alice's visited network consists of three instances, IBCF-1, IBCF-4 and IBCF-5, and controls the TrGW-1 entity.

The transit and roaming function (TRF) is located in the visited network of the caller (Alice). The TRF entity receives the initial request SIP INVITE from the S-CSCF entity of the home network of the caller and forwards the request to the home network of the called party (Bob).

Roaming interfaces are provided by:

– Ici interfaces between the IBCF-1 and-2 IBCF instances, between the IBCF-3 and IBCF-4 instances and between the IBCF-7 and IBCF-8 instances, for the SIP stream;

– Izi interface between the GW-3 and TrGW-4 entities, for the RTP stream.

The interfaces for the interconnection is provided by:

– Ici interface between the IBCF-5 and IBCF-6 instances for the SIP flow;

– Izi interface between the TrGW-1 and TrGW-3 entities for the RTP stream.

Figure 8.4 provides an example of SDP announcements containing the characteristics (IP addresses, port numbers and domain name) of the RTP stream.

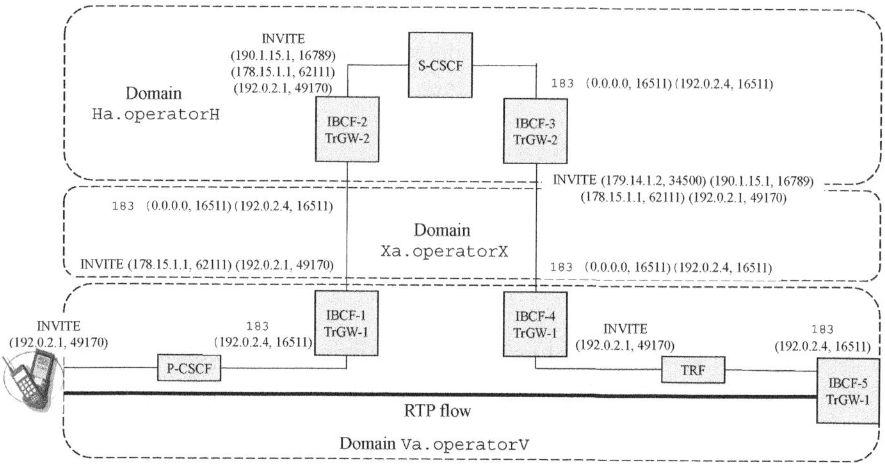

Figure 8.4. *SDP announcements: optimal routeing*

The SIP INVITE request generated by the various IBCF entities maintains in the SDP message the characteristics of RTP streams

(IP address, port number and domain name) of the previous announcements.

The IBCF-4 instance must detect from the SIP INVITE request received from the IBCF-3 instance that a looping from the home network has been achieved and must implement the OMR routeing for the RTP stream, to provide to the IBCF-5 instance the SDP message generated by Alice's UA entity.

The response SIP `183 Session Progress` from the IBCF-5 instance contains in the SDP message the characteristics of RTP streams provided by the TrGW-1 entity, which need to be transferred to Alice's UA entity.

The IBCF-4 instance replaces the IP address provided by the IBCF-5 instance by the undetermined IP address 0.0.0.0 as the IP address has no meaning in the home and transit networks.

The IBCF-4 instance, however, maintains in the SDP message the RTP stream characteristics provided by the IBCF-5 instance.

The undetermined IP address is deleted by IBCF-1 instance and the IP address provided by the IBCF-5 instance is restored.

8.2. Procedures

8.2.1. *Session establishment for nominal routeing*

8.2.1.1. *Originating side*

The session establishment procedure for the nominal routeing of the RTP streams, relating to the outgoing call, is described in Figure 8.5.

To simplify the presentation, responses `100 Trying` and `180 Ringing` and precondition mechanisms are not shown.

Table 8.1 summarizes the IP addresses and the port numbers of the RTP flows, established, on the one hand, between Alice's UA entity and the TrGW-1 entity, and on the other hand, between the TrGW-1 and TrGW-2 entities.

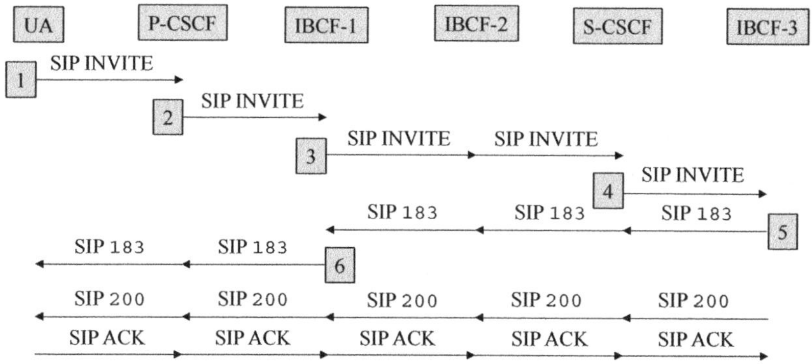

Figure 8.5. *Session establishment for nominal routeing originating side*

	Alice's UA	TrGW-1
IP address	192.0.2.1	192.0.2.2
Port number	49170	9452
	TrGW-1	TrGW-2
IP address	178.15.1.1	190.1.15.2
Port number	62111	12538

Table 8.1. *RTP flow characteristics in the case of nominal routeing originating side*

1) Alice's UA entity transmits to the P-CSCF entity the SIP INVITE request, whose SDP message contains the characteristics (IP address and port number) of the RTP stream.

```
INVITE SIP: tel:+4687197378; SIP/2.0
...
Route: <sip:pcscf1.visitedVA.net;lr>
...
Content-Type: application/sdp
Content-Length: (...)
...
c=IN IP4 192.0.2.1
m=audio 49170 RTP/AVP 96 97
...
```

2) The P-CSCF entity selects the IBCF entity of the visited network (IBCF-1 instance) and transfers the SIP INVITE request.

The P-CSCF entity removes its uniform resource identifier (URI) in the `Route` header and adds that of the S-CSCF entity of the home network, indicating in the `iotl` parameter the direction of request (`visitedA-homeA`).

```
INVITE SIP: tel:+4687197378; SIP/2.0
...
Route: <sip:scscf1.operatorHA.net;lr;iotl=visitedA-
homeA>
...
Content-Type: application/sdp
Content-Length: (...)
...
c=IN IP4 192.0.2.1
m=audio 49170 RTP/AVP 96 97
...
```

3) The IBCF-1 instance selects the IBCF entity of the home network (IBCF-2 instance) and transfers the SIP INVITE message, whose SDP message replaces the characteristics of RTP streams received from the Alice's UA entity by those provided by the TrGW-1 entity.

```
INVITE SIP: tel:+4687197378; SIP/2.0
...
Route: <sip:scscf1.operatorHA.net;lr;iotl=visitedA-
homeA>
...
Content-Type: application/sdp
Content-Length: (...)
...
c=IN IP4 178.15.1.1
m=audio 62111 RTP/AVP 96 97
...
```

The IBCF-2 instance forwards to the S-CSCF entity the SIP INVITE message without changing the SDP message.

4) The S-CSCF entity transfers to the IBCF-3 instance the SIP INVITE message having the following transactions:

– it withdraws its URI identity in the `route` header;

– it performs the ENUM resolution on the URI identity of the destination to which it adds the `iotl` parameter indicating the direction of the request (`homeA-homeB`).

```
INVITE
<sip:+46107197378@operatorY;user=phone;iotl=homeA-homeB>
SIP/2.0
...
Content-Type: application/sdp
Content-Length: (...)
...
c=IN IP4 178.15.1.1
m=audio 62111 RTP/AVP 96 97
...
```

When the IBCF-3 instance has received the SIP INVITE request, it generates a new SIP INVITE request to Bob's home network.

5) On receipt of the SIP `183 Session Progress` message from Bob's home network, the IBCF-3 instance generates the SIP `183 Session Progress` message to the S-CSCF entity whose associated SDP message contains the characteristics of the flow RTP communicated by the TrGW-2 entity.

```
SIP/2.0 183 Session Progress
...
Content-Type: application/sdp
Content-Length: (...)
...
c=IN IP4 190.1.15.2
m=audio 12538 RTP/AVP 97 98
...
```

The SIP `183 Session Progress` message is forwarded without changing the SDP message to the IBCF-1 instance.

6) The IBCF-1 instance forwards to the P-CSCF entity the SIP `183 Session Progress` message, whose SDP message replaces the RTP stream characteristics received from the IBCF-3 instance with those provided by the entity TrGW-1.

```
SIP/2.0 183 Session Progress
...
Content-Type: application/sdp
Content-Length: (...)
...
c=IN IP4 192.0.2.2
m=audio 9452 RTP/AVP 97 98
...
```

The SIP `183 Session Progress` message is forwarded without changing the SDP message to Alice's UA entity.

8.2.1.2. *Terminating side*

The session establishment procedure for the nominal routeing of the RTP streams, relating to the outgoing call, is described in Figure 8.6.

Table 8.2 summarizes the IP addresses and the port numbers of the RTP streams, established, on the one hand, between Bob's UA entity and TrGW-4 entity, and, on the other hand, between TrGW-3 and TrGW-4 entities.

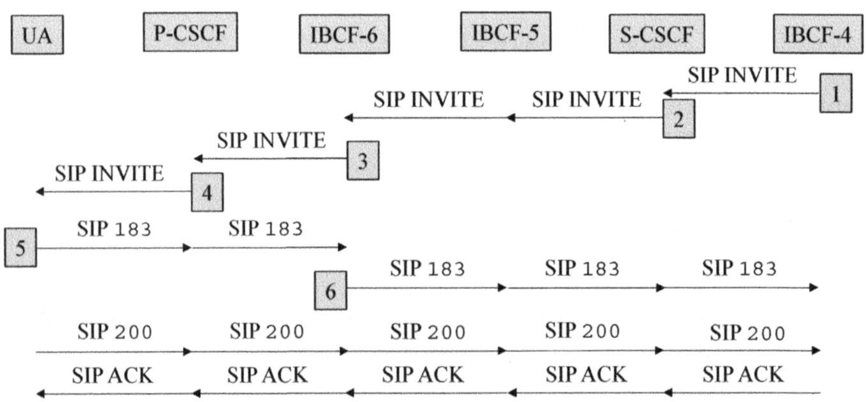

Figure 8.6. *Session establishment for nominal routeing terminating side*

	Bob's UA	TrGW-4
IP address	193.0.2.1	193.0.2.2
Port number	49170	9452
	TrGW-4	TrGW-3
IP address	179.15.1.1	191.1.15.2
Port number	62111	12538

Table 8.2. *RTP flow characteristics in the case of nominal routeing terminating side*

1) Upon receipt of the SIP INVITE request from Alice's home network, the IBCF-4 instance generates the SIP INVITE message to the S-CSCF entity, whose SDP message contains the RTP stream characteristics provided by the TrGW-3 entity.

```
INVITE
<sip:+46107197378@operatorY;user=phone;iotl=homeA-homeB>
SIP/2.0
...
Content-Type: application/sdp
Content-Length: (...)
...
c=IN IP4 191.1.15.2
m=audio 12538 RTP/AVP 96 97
...
```

2) The S-CSCF entity inserts Bob's IP address instead of the phone number into the SIP URI identity of the SIP INVITE request.

The S-CSCF entity adds the `Route` header containing the URI identity of the P-CSCF entity and `iotl` parameter indicating the direction of the request (`homeB-visitedB`).

The IBCF-5 instance forwards to the IBCF (IBCF-6 instance) entity the SIP INVITE message without changing the SDP message.

```
INVITE <sip:193.0.2.1@operatorY> SIP/2.0
...
Route: <sip:pcscf1.visitedVB.net;lr;iotl= homeB-visitedB>
...
Content-Type: application/sdp
Content-Length: (...)
...
c=IN IP4 191.1.15.2
m=audio 12538 RTP/AVP 96 97
...
```

3) The IBCF-6 instance transfers to the P-CSCF entity the SIP INVITE message, whose SDP message replaces the RTP stream characteristics received from TrGW-3 entity with that communicated by the TrGW-4 entity.

```
INVITE <sip:193.0.2.1@operatorY> SIP/2.0
...
Route: <sip:pcscf1.visitedVB.net;lr;iotl= homeB-visitedB>
...
Content-Type: application/sdp
Content-Length: (...)
...
c=IN IP4 193.0.2.2
m=audio 9452 RTP/AVP 96 97
...
```

4) The P-CSCF entity removes its identity from the `Route` header and transfers the SIP INVITE message to Bob's UA entity.

```
INVITE <sip:193.0.2.1@operatorY> SIP/2.0
...
Content-Type: application/sdp
Content-Length: (...)
...
c=IN IP4 193.0.2.2
m=audio 9452 RTP/AVP 96 97
...
```

5) Alice's UA entity generates the SIP `183 Session Progress` message to the P-CSCF entity, whose associated SDP message contains the RTP stream characteristics.

```
SIP/2.0 183 Session Progress
...
Content-Type: application/sdp
Content-Length: (...)
...
c=IN IP4 193.0.2.1
m=audio 59170 RTP/AVP 97 98
...
```

The SIP `183 Session Progress` message is forwarded without changing the SDP message to the IBCF-6 instance.

6) The IBCF-6 instance forwards to the IBCF-5 instance SIP `183 Session Progress` message, whose SDP message replaces the RTP stream characteristics received from the Bob's UA entity with those provided by the TrGW-4 entity.

```
SIP/2.0 183 Session Progress
...
Content-Type: application/sdp
Content-Length: (...)
...
c=IN IP4 179.15.1.1
m=audio 62111 RTP/AVP 97 98
...
```

The SIP `183 Session Progress` message is forwarded without changing the SDP message to the IBCF-4 instance.

8.2.2. *Session establishment for optimal routeing*

The session establishment procedure for the OMR routeing of the RTP streams, relating to the outgoing call, is shown in Figure 8.7.

1) Alice's UA entity transmits to the P-CSCF entity the SIP INVITE request, whose SDP message contains the characteristics (IP address and port number) of the RTP stream.

```
INVITE SIP: tel:+4687197378; SIP/2.0
...
Route: <sip:pcscf1.visitedVA.net;lr>
...
Content-Type: application/sdp
Content-Length: (...)
...
c=IN IP4 192.0.2.1
m=audio 49170 RTP/AVP 96 97
...
```

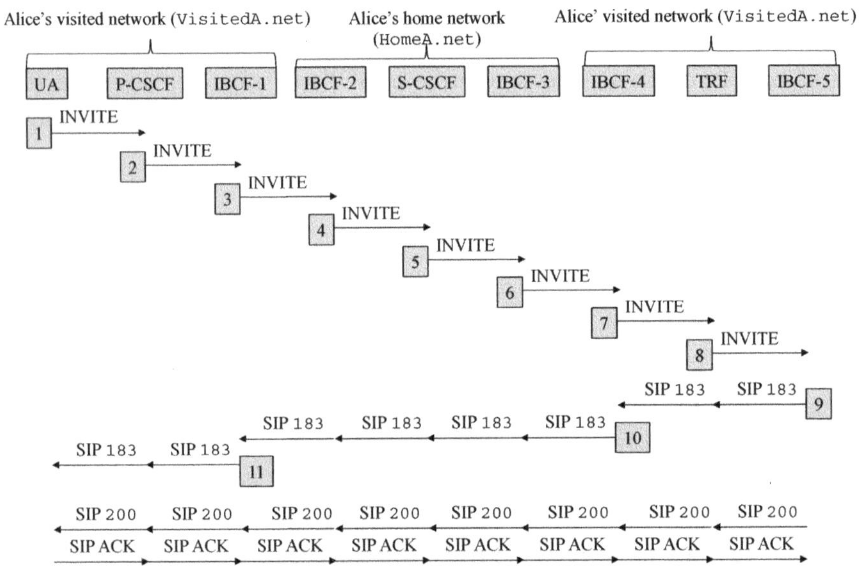

Figure 8.7. *Session establishment for optimal routeing originating side*

2) The P-CSCF entity removes its URI identity in the `Route` header and adds that of the S-CSCF entity of the home network, indicating in the `iotl` parameter the direction of request (`visitedA-homeA`).

The P-CSCF entity selects the IBCF entity of the visited network (IBCF-1 instance) and transfers the SIP INVITE request, whose `Feature-Caps` header contains the URI identity of the TRF entity.

```
INVITE SIP: tel:+4687197378; SIP/2.0
...
Route: <sip:scscf1.operatorHA.net;lr;iotl=visitedA-
homeA>
Feature-
Caps:*;+g.3gpp.trf="<sip:trf1.visitedV.net;iotl=homeA-
visitedA>"
...
Content-Type: application/sdp
Content-Length: (...)
...
c=IN IP4 192.0.2.1
m=audio 49170 RTP/AVP 96 97
...
```

3) The IBCF-1 instance selects the IBCF entity of the home network (IBCF-2 instance) and transfers the SIP INVITE message, whose SDP message replaces the RTP stream characteristics received from Alice's UA entity with those provided by the TrGW-1 entity.

The SIP message also contains the instances 1 and 2 of the `a=visited-realm` populated with the following information:

– the domain name to which Alice's UA entity is connected (`Va.operatorV.net`), the IP address and the RTP stream port number of Alice's UA entity;

– the domain name of the network to which the IBCF-1 instance is connected (`Xa1.operatorX.net`), the IP address and the port number of the RTP stream provided by the TrGW-1 entity for the specified domain name.

```
INVITE SIP: tel:+4687197378; SIP/2.0
...
Route: <sip:scscf1.operatorH.net;lr;iotl=visitedA-homeA>
...
Content-Type: application/sdp
Content-Length: (...)
...
c=IN IP4 178.15.1.1
m=audio 62111 RTP/AVP 96 97
...
a=visited-realm:1 Va.operatorV.net IN IP4 192.0.2.1
49170
a=visited-realm:2 Xa1.operatorX.net IN IP4 178.15.1.1
62111
...
```

4) The IBCF-2 instance forwards to the S-CSCF entity the SIP INVITE message, whose SDP message replaces the RTP stream characteristics received from the IBCF-1 instance with those provided by the TrGW-2 entity.

The SIP message adds the instance 3 of the header `a=visited-realm` populated with the domain name of the network to which the instance IBCF-2 is connected (`Ha.operatorH.net`), the IP address and the port number of the RTP stream provided by the TrGW-2 entity for the specified domain name.

```
INVITE SIP: tel:+4687197378; SIP/2.0
...
Route: <sip:scscf1.operatorH.net;lr;iotl=visitedA-homeA>
...
Content-Type: application/sdp
Content-Length: (...)...
c=IN IP4 190.1.15.1
m=audio 16789 RTP/AVP 96 97
...
a=visited-realm:1 Va.operatorV.net IN IP4 192.0.2.1
49170
a=visited-realm:2 Xa1.operatorH.net IN IP4 178.15.1.1
62111
a=visited-realm:3 Ha.operatorH.net IN IP4 190.1.15.1
16789
...
```

5) The S-CSCF entity transfers to the IBCF-3 instance the SIP INVITE message whose header `Feature-Caps` indicates that the loop to Alice's visited network is activated.

The S-CSCF entity withdraws its URI identity in the `Route` header and adds that the TRF entity of the visited network, indicating in the `iotl` parameter the direction of the request (`homeA-visitedA`).

```
INVITE SIP: tel:+4687197378; SIP/2.0
...
Route:<sip:trf1.visitedV.net;lr;iotl=homeA-visitedA>
Feature-Caps:*;+g.3gpp.loopback=<"homenetwork_A">
...
Content-Type: application/sdp
Content-Length: (...)
...
```

```
c=IN IP4 190.1.15.1
m=audio 16789 RTP/AVP 96 97
...
a=visited-realm:1 Va.operatorV.net IN IP4 192.0.2.1
49170
a=visited-realm:2 Xa1.operatorH.net IN IP4 178.15.1.1
62111
a=visited-realm:3 Ha.operatorH.net IN IP4 190.1.15.1
16789
...
```

6) The IBCF-3 instance transfers to the IBCF entity of the visited network (IBCF-4 instance) the SIP INVITE message, whose SDP message replaces the RTP stream characteristics received from the IBCF-2 instance with those reported by the TrGW-2 entity.

The SIP message adds the instance 4 of the header `a=visited-realm` populated with the domain name of the network to which the IBCF-3 instance is connected (`Xa2.operatorX.net`), the IP address and the port number of the RTP, stream provided by the TrGW-2 entity for the specified domain name.

```
INVITE SIP: tel:+4687197378; SIP/2.0
...
Route:<sip:trf1.visitedV.net;lr;iotl=homeA-visitedA>
Feature-Caps:*;+g.3gpp.loopback=<"homenetwork-A">
...
Content-Type: application/sdp
Content-Length: (...)
...
c=IN IP4 179.14.1.2
m=audio 34500 RTP/AVP 96 97
...
a=visited-realm:1 Va.operatorV.net IN IP4 192.0.2.1
49170
a=visited-realm:2 Xa1.operatorX.net IN IP4 178.15.1.1
62111
a=visited-realm:3 Ha.operatorH.net IN IP4 190.1.15.1
16789
a=visited-realm:4 Xa2.operatorX.net IN IP4 179.14.1.2
34500
...
```

7) Upon receipt of the SIP INVITE message, the IBCF-4 instance detects that the case of optimal routeing is possible.

The IBCF-4 instance transfers to the TRF entity the SIP INVITE message, whose SDP message replaces the RTP stream characteristics of the IBCF-3 instance with those of the instance 1 of the header `a=visited-realm`.

Instances 2 to 4 of the header `a=visited-realm` are deleted.

```
INVITE SIP: tel:+4687197378; SIP/2.0
Route:<sip:trf1.visitedV.net;lr;iotl=homeA-visitedA>
Feature-Caps:*;+g.3gpp.loopback=<"homenetwork-A">
...
Content-Type: application/sdp
Content-Length: (...)
...
c=IN IP4 192.0.2.1
m=audio 49170 RTP/AVP 96 97
...
a=visited-realm:1 Va.operatorV.net IN IP4 192.0.2.1
49170
...
```

8) The TRF entity performs ENUM resolution on the URI identity of the destination to which it adds the `iotl` parameter indicating the direction of the request (`visitedA-homeB`), withdraws its URI identity in the `Route` header and transfers the SIP INVITE request to the IBCF-5 instance.

```
INVITE SIP:
<sip:+46107197378@operatorY;user=phone;iotl=visitedA-
homeB> SIP/2.0
...
Feature-Caps:*;+g.3gpp.loopback=<"homenetwork-A">
...
Content-Type: application/sdp
Content-Length: (...)
...
c=IN IP4 192.0.2.1
m=audio 49170 RTP/AVP 96 97
...
a=visited-realm:1 Va.operatorV.net IN IP4 192.0.2.1
49170
...
```

When the IBCF-5 instance received the SIP INVITE request, it generates a new request to the SIP INVITE Bob's home network.

9) Upon receipt of the SIP `183 Session Progress` message from Bob's home network, the IBCF-5 instance generates to the TRF entity, the SIP `183 Session Progress` message, whose associated SDP message contains RTP stream characteristics reported by the TrGW-1 entity.

The SIP message also contains the instance 1 of the header `a=visited-realm` populated with the domain name to which the IBCF-5 instance is connected the IP address and the RTP stream port number of the TrGW-1 entity.

```
SIP/2.0 183 Session Progress
...
Content-Type: application/sdp
Content-Length: (...)
...
c=IN IP4 192.0.2.4
m=audio 16511 RTP/AVP 97 98
...
a=visited-realm:1 Va.operatorV.net IN IP4 192.0.2.4 16511
...
```

The SIP `183 Session Progress` message is forwarded without changing the SDP message to the IBCF-4 instance.

10) The IBCF-4 instance transfers to the IBCF entity of the visited network (IBCF-3 instance), the SIP `183 Session Progress` message whose SDP message replaces the IP address of the RTP stream of the IBCF-5 instance by 0.0.0.0 value.

```
SIP/2.0 183 Session Progress
...
Content-Type: application/sdp
Content-Length: (...)
...
c=IN IP4 0.0.0.0
m=audio 16511 RTP/AVP 97 98
...
a=visited-realm:1 Va.operatorV.net IN IP4 192.0.2.4 16511
...
```

The SIP `183 Session Progress` message is forwarded without changing the SDP message to the IBCF-1 instance.

11) The IBCF-1 instance forwards to the P-CSCF entity the SIP `183 Session Progress` message, whose SDP message replaces the RTP stream characteristics received from the IBCF-4 instance with those of the instance 1 of the header `a=visited-realm` and removes it.

The SIP `183 Session Progress` message is forwarded without changing the SDP message to Alice's UA entity.

```
SIP/2.0 183 Session Progress
...
Content-Type: application/sdp
Content-Length: (...)
...
c=IN IP4 192.0.2.4
m=audio 16511 RTP/AVP 97 98
...
```

9

Service Centralization and Continuity

9.1. ICS function

IMS centralized services (ICS) allow for centralizing IMS services regardless of whether the mode of the mobile network is circuit-switched (CS) or packet-switched (PS).

The role of the network in CS mode becomes equivalent to that of the network in PS mode: it is restricted to the construction of bearers which handle telephone signaling and voice or conversational video.

9.1.1. *Functional architecture*

The functional architecture of the ICS function is described in Figure 9.1, for the case where the mobile-services switching centre (MSC) server and the user equipment (UE) implement ICS.

The mobile attached to the network in CS mode can use the Gm interface for session initiation protocol (SIP) if both CS and PS modes are available simultaneously.

The mobile attached to the network in CS mode can use the Ut interface to configure its services to the telephony application server (TAS) if both CS and PS modes are available simultaneously.

The ICS function introduces a new entity in the IMS network, the service centralization and continuity application server (SCC AS).

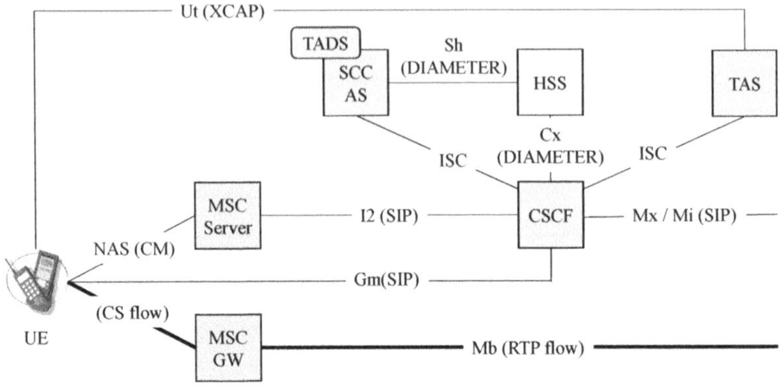

Figure 9.1. *MSC server and UE implementing ICS function functional architecture*

The SCC AS entity combines the SIP signaling arising from the mobile (via the Gm interface) and from the MSC server entity (via the I2 interface) and acts as a back-to-back user agent (B2BUA), providing an anchor point for incoming and outgoing calls:

– in the case of an outgoing call, it ends the dialogue with the mobile and the MSC server entity and initiates a new dialogue with the downstream entity. The SCC AS entity is the first application server invoked by the serving call session control function (S-CSCF).

– in the case of an incoming call, it ends the dialogue with the upstream entity and initiates a dialogue with the MSC server entity and the mobile. The SCC AS entity is the latest application server invoked by the S-CSCF entity.

The SCC AS entity provided the terminating access domain selection (T-ADS) that can route an incoming call to the appropriate mobile network: 2G or 3G network in CS mode, 3G or 4G network in PS mode (Figure 9.2).

The SCC AS entity queries the home subscriber server (HSS) to obtain the address of the mobility management entity (MME) and the service GPRS support node (SGSN).

The HSS entity asks the SGSN and MME entities to recover network and mobile capacities.

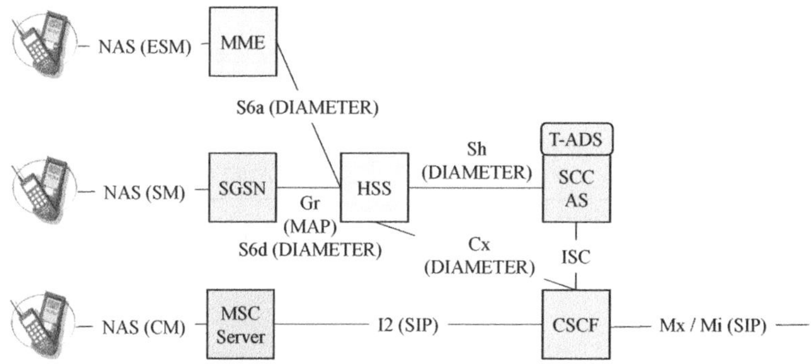

Figure 9.2. *Functional architecture for TADS*

The MSC server entity ensures registering the mobile from the SCC AS and S-CSCF entities.

The MSC server entity provides the translation of the telephone signaling for the establishment of a communication:

– call management (CM), a part of non-access stratum (NAS), at CS side;

– SIP signaling, at IMS side via the I2 interface.

The MSC server entity provides the control of the MSC GW entity for the establishment of the CS bearer, on one side, and the real-time transport protocol (RTP) stream on the side of the IP network.

The functional architecture of the ICS function is described in Figure 9.3 in the case where the MSC server entity and the mobile do not implement the ICS service.

When the MSC server entity receives the call for the establishment of the communication, it recovers from the SCC AS entity, via camel application part (CAP), the IP multimedia routing number (IMRN).

The MSC server entity routes the call to the media gateway control function (MGCF) using the IMRN number to reach the SCC AS entity.

The MGCF entity translates the telephone signaling:

– ISDN user part (ISUP) at CS network side;

– SIP at IMS network side.

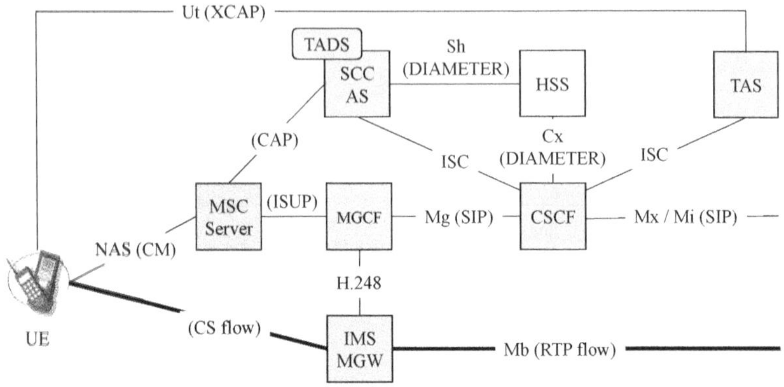

Figure 9.3. *MSC server and UE not implementing ICS function functional architecture*

The MGCF entity provides the control of the IMS GW entity for the establishment of the CS bearer on one side and the RTP stream on the IP network side.

The B2BUA function of the SCC AS entity allows to end the dialogue with the MGCF entity and to start a new dialogue with the called party.

9.1.2. *Procedures*

9.1.2.1. *Registration*

The registration of the mobile to the IMS network deploys the general procedure if the mobile implements the ICS function.

In the case where the mobile and the MSC server entity do not implement the ICS function, the mobile is not registered to the IMS network.

The registration procedure of the mobile to the IMS network is described in Figure 9.4, where:

– the MSC server entity implements ICS function;

– the mobile does not implement ICS function.

1) After the attachment of the mobile, the MSC server entity transmits the SIP REGISTER request to the I-CSCF (interrogating-CSCF) entity.

The SIP URI (Uniform Resource Identifier) of the REGISTER request is derived from the mobile country code (MCC) and the mobile network code (MNC).

The temporary public identity of the headers `From` and `To` is derived from the international mobile subscriber identity (IMSI).

The private identity in the `Authorization` header is also derived from the IMSI private identity.

The `Contact` header contains the parameter `+ g.3gpp.ics` indicating that the MSC server entity implements the ICS service.

```
REGISTER sip:ics.mnc01.mcc208.3gppnetwork.org SIP/2.0
...
From:
<sip:208109999999999@ics.mnc01.mcc208.3gppnetwork.org>;
tag=4fa3
To:
<sip:208109999999999@ics.mnc01.mcc208.3gppnetwork.org>
Contact: <sip:[5555::aaa:bbb:ccc:ddd]>;expires=600000;
+g.3gpp.ics="server"
Authorization: Digest username="
208109999999999@ics.mnc01.mcc208.3gppnetwork.org ",
realm=" ics.mnc01.mcc208.3gppnetwork.org ", nonce="",
integrity-protected="auth-done", uri="sip:
ics.mnc01.mcc208.3gppnetwork.org ", response=""
...
```

2) The I-CSCF entity sends to the HSS entity the message DIAMETER user-authorization-request (UAR) to retrieve the list of the S-CSCF entities that can be assigned to the UA entity.

3) The I-CSCF entity performs the selection of an S-CSCF entity to which it forwards the REGISTER request, from the list of the S-CSCF entities received in the message DIAMETER user-authorization-answer (UAA).

4) The I-CSCF entity replaces the initial URI identity (`sip: ics.mnc01.mcc208.3gppnetwork.org`) by that of the S-CSCF entity (`sip:scscf.HomeA.net or sip:scscf.HomeB.net`) and transmits the SIP REGISTER request to the selected S-CSCF entity.

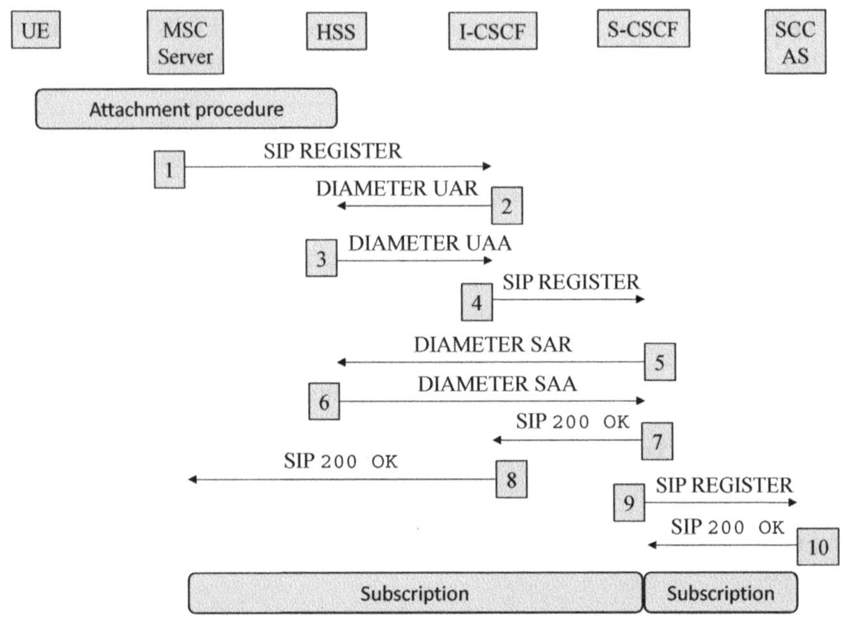

Figure 9.4. *Mobile registration to IMS network with ICS function*

5) As the mobile has been authenticated by the MSC server entity, the S-CSCF entity sends to the HSS entity the message DIAMETER server-assignment-request (SAR) to retrieve the mobile profile.

6) The HSS entity transmits the profile of the mobile in the message DIAMETER server-assignment-answer (SAA).

7), 8) The SIP `200 OK` response follows the reverse route taken by the REGISTER request.

9) The S-CSCF entity registers the mobile to the SCC AS entity.

10) The SCC AS entity responds with SIP `200 OK` message to the SIP REGISTER request.

After the registration procedure, the MSC server and SCC AS entities deploy the general procedure for subscription to the mobile registration events.

9.1.2.2. *Session establishment at originating side*

The procedure to establish the session for the outgoing call is described in Figure 9.5, in the case where the MSC server entity and the mobile implement ICS function.

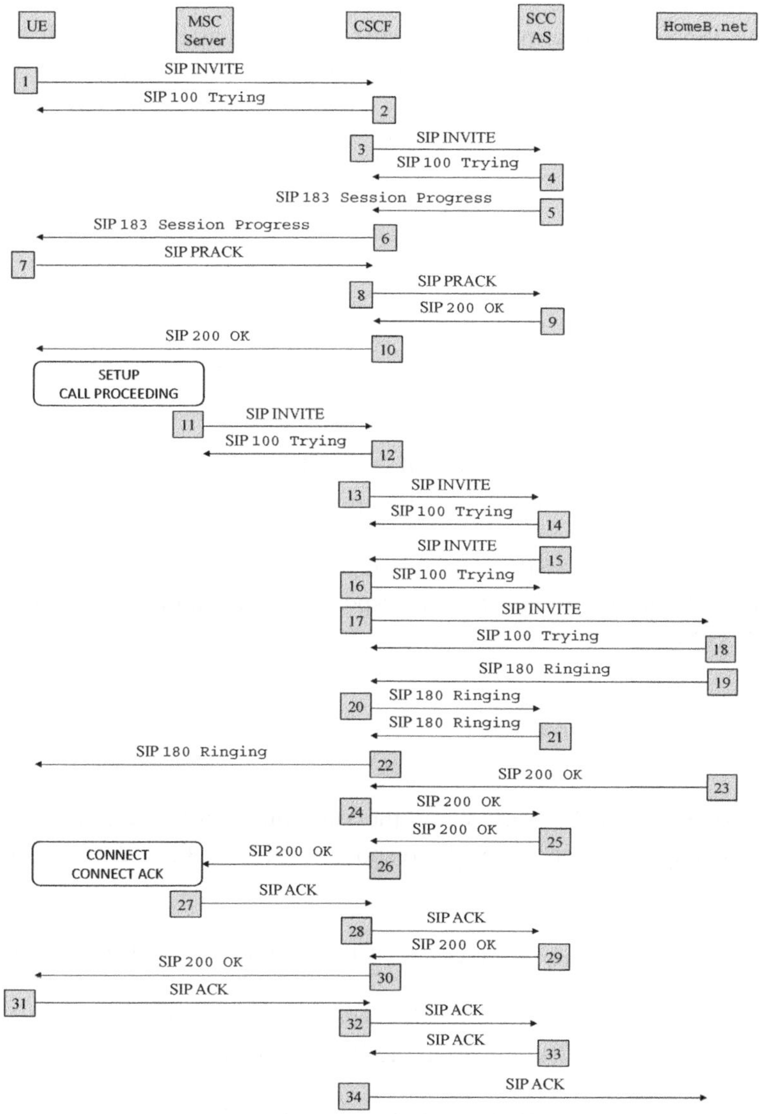

Figure 9.5. *Session establishment at originating side MSC server and UE implementing ICS function*

1), 3) Alice's UA entity initializes the service control by sending the SIP INVITE message to the SCC AS entity.

The URI identity of the request is the SIP URI or TEL URI identity of the called party (Bob).

SDP (Session Description Protocol) messages associated with the SIP INVITE request indicates that the session is established from a CS-mode network.

```
INVITE tel:+1-212-555-2222 SIP/2.0
...
c=PSTN - -
t=0 0
m=audio 9 PSTN -
a=setup:active
a=connection:new
a=cs-correlation:callerid:+358504821437
...
```

2), 4) Each entity responds with the SIP `100 Trying` response that allows blocking the retransmission timer of the SIP INVITE request.

5), 6) The response SIP `183 Session in Progress` received from the SCC AS entity provides a SDP response that contains the public service identity (PSI).

```
...
c=PSTN E164 +12125556666
...
```

7), 8) Alice's UA entity sends the subsequent SIP PRACK request to acknowledge the provisional response SIP `183 Session in Progress`.

9), 10) The SIP `200 OK` message is the response to the PRACK request.

Alice's mobile proceeds with the establishment of the bearer in the CS-mode network (SETUP) indicating the PSI identity as destination.

11), 13) The MSC server entity generates the SIP INVITE request to the identity PSI.

The header `P-Asserted-Identity` contains the phone number of the caller.

The header `Accept-Contact` indicates that the MSC server entity supports supplementary telephone services.

```
INVITE tel:+1-212-555-6666 SIP/2.0
...
P-Asserted-Identity: <tel:+358-50-4821437>
Accept-Contact: *;+g.3gpp.icsi-ref="urn%3Aurn-7%3gpp-service.ims.icsi.mmtel"
...
```

The SIP INVITE request contains an SDP offer that contains the characteristics of the audio stream provided by the MSC GW entity.

12), 14) Each entity responds with the SIP `100 Trying` response that allows for blocking the retransmission timer of the SIP INVITE request.

15), 17) The SCC AS entity generates the SIP INVITE request to Bob located in the domain `HomeB.net`.

The SCC AS entity replaces the PSI identity by the TEL URI identity of Bob that was recorded during the first SIP INVITE message.

```
INVITE tel:+1-212-555-2222 SIP/2.0
...
```

16), 18) Each entity responds with the SIP `100 Trying` response that allows for blocking the retransmission timer of the SIP INVITE request.

19), 20) The SCC AS entity receives from the domain `HomeB.net` the SIP `180 Ringing` message indicating that Bob’s phone is ringing.

21), 22) The SCC AS entity responds with the SIP `180 Ringing` message to the INVITE request received from Alice’s UA entity.

23), 24) The SCC AS entity receives from the domain `HomeB.net` the SIP `200 OK` message indicating that Bob picked up the phone.

The SIP `200 OK` message contains the SDP message with the characteristics of the audio stream.

25), 26) The SCC AS entity responds to the SIP INVITE request of the MSC server entity with the SIP `200 OK` message in which the SCC AS entity forwards the SDP message received from the domain `HomeB.net`.

Upon receipt of the SIP `200 OK` message, the MSC server entity confirms the session establishment (CONNECT) to Alice's mobile and transfers the characteristics of audio stream to the MSC GW entity.

27), 28) The MSC server entity acknowledges the SIP `200 OK` response by the subsequent SIP ACK request.

29), 30) The SCC AS entity responds to the SIP INVITE request of Alice's UA entity with the SIP `200 OK` message.

31), 32) Alice's UA entity acknowledges the SIP `200 OK` response by the subsequent SIP ACK request.

33), 34) The SCC AS entity acknowledges the SIP `200 OK` response received from the area `HomeB.net` by the subsequent SIP ACK request.

9.1.2.3. *Session establishment at terminating side*

The procedure to establish the session for the incoming call is illustrated in Figure 9.6, in the case where the MSC server entity and the mobile implement ICS function.

1), 3) The SCC AS entity receives the SIP INVITE request from the domain `HomeA.net`. The translation of TEL URI identity in SIP URI has been issued within the network `HomeA.net`.

```
INVITE sip:bob@HomeB.net SIP/2.0
...
```

The SDP message associated with the SIP INVITE request contains the characteristics of the audio stream.

2), 4) Each entity responds with the SIP `100 Trying` response that allows blocking the retransmission timer of the SIP INVITE request.

5) The SCC AS entity determines the capacity and the location of the mobile through the T-ADS function and generates a SIP INVITE request to Bob's UA entity.

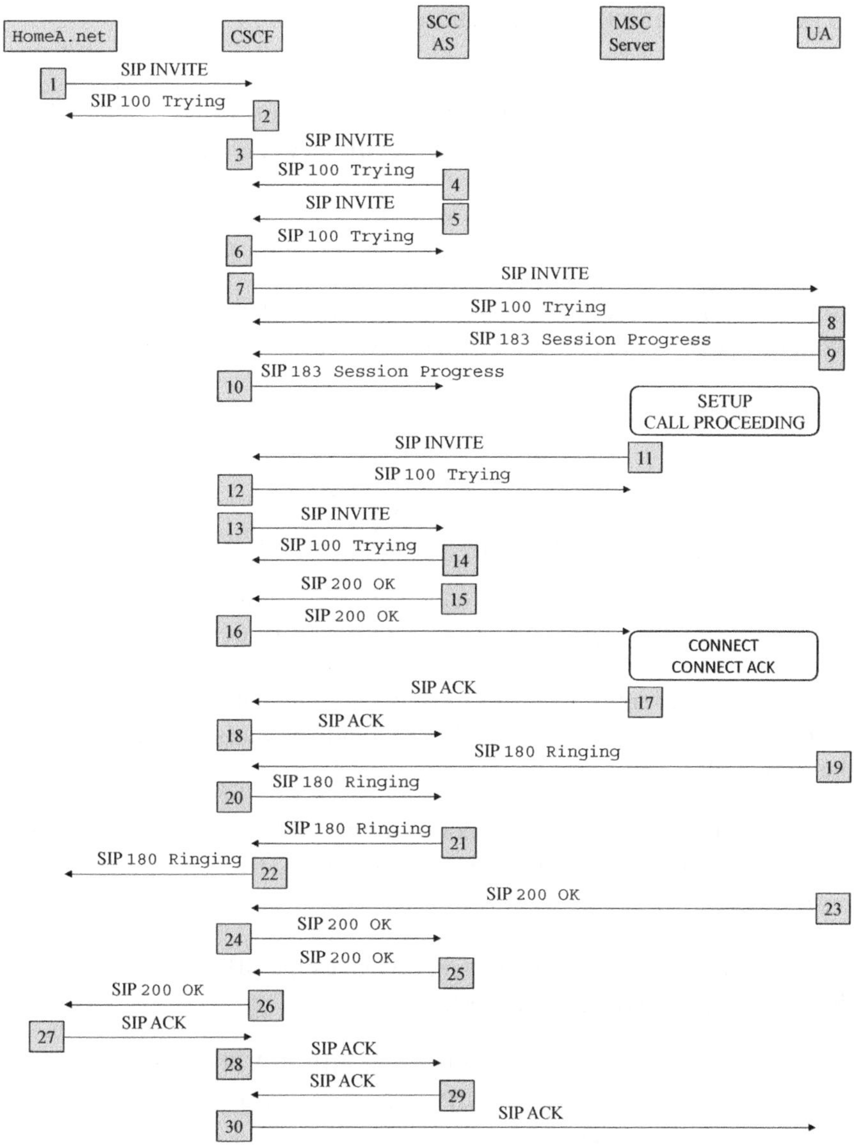

Figure 9.6. *Session establishment at terminating side MSC server and UE implementing ICS function*

7) The S-CSCF entity replaces the SIP URI identity by Bob's telephone number associated with the identity of the MSC Server which recorded Bob.

```
INVITE sip:12125552222@msc.HomeB.net SIP/2.0
...
```

6), 8) Each entity responds with SIP `100 Trying` response that allows for blocking the retransmission timer of the SIP INVITE request.

9), 10) Bob's UA entity responds with SIP `183 Session Progress` message containing the SDP message indicating that the CS bearer is used.

Bob's mobile proceeds with the establishment of the bearer in the CS network (SETUP), indicating the PSI identity as destination.

11), 13) The MSC Server entity generates a SIP INVITE request to the PSI identity.

The SIP INVITE request contains an SDP offer that contains the characteristics of the audio stream provided by the MSC GW entity.

12), 14) Each entity responds with SIP `100 Trying` response that allows for blocking the retransmission timer of the SIP INVITE request.

The MSC server entity indicates to Bob's mobile that the media in the CS network has been established (CONNECT) and forwards the SDP message to the MSC GW entity.

15), 16) The SCC AS entity responds to the MSC server entity with SIP `200 OK` message containing the SDP message received from the domain `HomeA.net`.

17), 18) The MSC server entity acknowledges the SIP `200 OK` response with the subsequent SIP ACK request.

19), 20) Bob's UA entity responds to the SIP INVITE request received from the SCC AS entity with the SIP `180 Ringing` message to indicate that Bob's phone is ringing.

23), 24) Bob's UA entity responds to the SIP INVITE request received from the SCC AS entity with SIP `200 OK` message to indicate that Bob picked up the phone.

25), 26) The SCC AS entity responds to the SIP INVITE request received from the domain `HomeA.net` with the SIP `200 OK` message to indicate that Bob picked up the phone.

27), 28) The domain `HomeA.net` acknowledges the SIP `200 OK` response with the subsequent SIP ACK request.

29), 30) The SCC AS entity acknowledges the SIP `200 OK` response with the subsequent SIP ACK request.

9.2. e-SRVCC function

9.2.1. *Functional architecture*

9.2.1.1. *Architecture for basic call*

When the mobile has established a call on the 4G evolved packet system (EPS), it is necessary to be able to transfer to the 3G universal mobile telecommunications system (UMTS) or 2G global system for mobile (GSM), if loss of coverage occurs on the 4G EPS network.

The enhanced single radio voice call continuity (e-SRVCC) ensures continuity of service when the mobile moves from one network in PS mode to a network in CS mode.

The functional architecture of the e-SRVCC function is described in Figure 9.7, for the SIP flow and in Figure 9.8 for the RTP stream.

To ensure continuity of service, e-SRVCC function introduces two anchor points in the IMS network:

– access transfer control function (ATCF) which provides the anchor point for SIP signaling. The ATCF entity is inserted into the path of SIP signaling between control session CSCF entities, on one hand, the P-CSCF (Proxy-CSCF) entity while, on the other hand, the I-CSCF or the S-CSCF entity;

– access gateway transfer (ATGW) that provides the anchor point for the RTP stream.

The ATCF entity is located in the visited IMS network in the case of roaming to hide to the nominal IMS network and to the interconnected IMS network that the SIP flow has changed IP address, given the move from the PS mode to the CS mode.

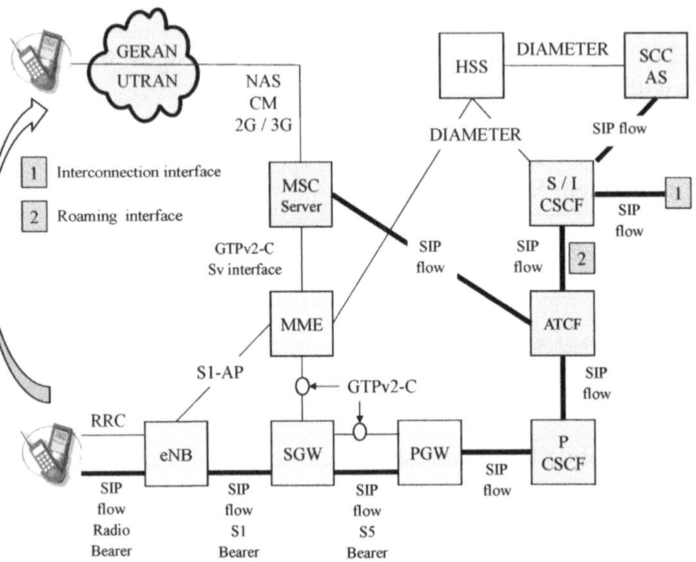

Figure 9.7. *Functional architecture for basic call control plane*

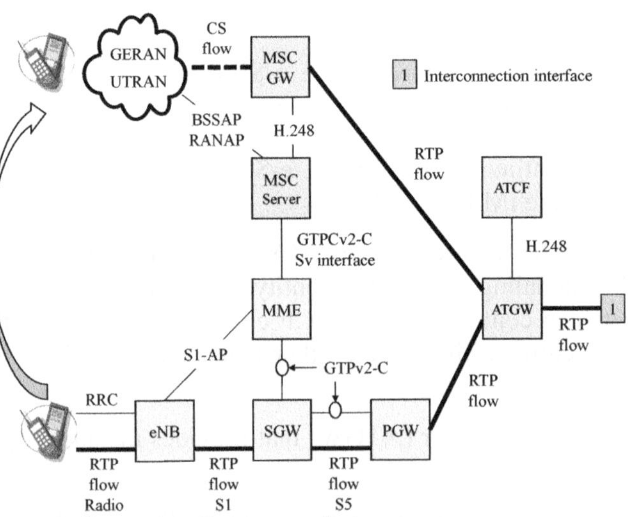

Figure 9.8. *Functional architecture for basic call traffic plane*

The ATGW entity is located in the visited IMS network too in the case of roaming to hide from the interconnected IMS network that the RTP stream has changed its IP address, given the move from the PS mode to the CS mode.

The ATGW entity is controlled by the ATCF entity via the H.248 protocol.

The ATGW entity may also perform the transcoding of voice if the codecs used, firstly, in the EPS network and, secondly, in GSM or UMTS networks, are different.

Figure 9.9 provides an example of the constitution of the RTP stream between Alice's UA in the domain `HomeA.net`, Bob's UA in the domain `HomeB.net` and ATGW entities.

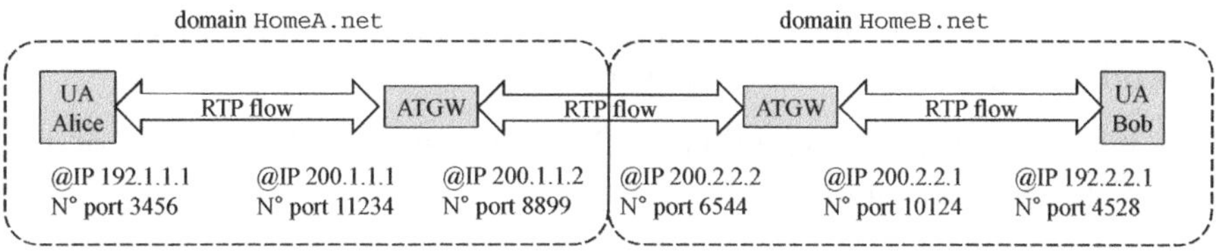

Figure 9.9. *RTP flow characteristics for service continuity*

Figure 9.10 provides an example of the constitution of the RTP stream between MSC GW entity in the domain `HomeA.net`, Bob's UA in the domain `HomeB.net` ATGW entities, after the PS-CS inter-system handover for Alice's mobile.

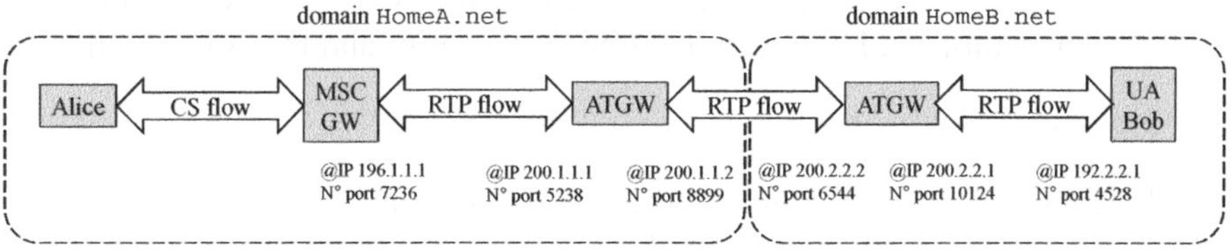

Figure 9.10. *RTP flow characteristics at the end of PS-CS inter-system handover*

The SCC AS entity implements the mechanisms for the control of the registration of the e-SRVCC function and for the transfer of the session at the PS-CS inter-system handover.

The MME entity of the EPS network is affected by the e-SRVCC function by performing the following functions:

– it separates the bearer dedicated to the voice from the other media carrying no voice;

– it initializes, via the Sv interface, the e-SRVCC procedure for voice handover to the target cell of the GSM or UMTS network;

– it coordinates the handover from the PS mode to the CS mode for voice and possibly the handover of the PS mode to the PS mode for the other streams.

The MSC server entity of the 2G GSM or 3G UMTS network is also affected by the e-SRVCC function by performing the following functions:

– it ensures the availability of resources in the 2G GSM or 3G UMTS network before executing the handover;

– it coordinates the execution of the handover and the transfer of the telephone communication.

The transfer of the telephone signaling involves transferring, on one hand, SIP signaling exchanged between the mobile and the ATCF entity, while on the other hand, the signaling consisting of call management exchanged between the mobile and the MSC server and SIP exchanged between MSC server and ATCF entity.

The voice transfer involves transferring, on one hand, the RTP stream established between the mobile and the ATGW entity, while on the other hand, the CS bearer between the mobile and the MSC GW entity and the RTP stream established between the MSC GW entity and the ATGW entity.

9.2.1.2. *Architecture for emergency call*

The functional architecture of the e-SRVCC function in the case of an emergency call is given in Figure 9.11.

The emergency access transfer function (EATF) ensures continuity of service when the mobile performs the PS-CS inter-system handover.

The EATF entity constitutes the anchor point of the SIP signaling and acts as a B2BUA entity.

In the case of roaming, the E-CSCF (Emergency CSCF) and EATF entities are located in the visited network.

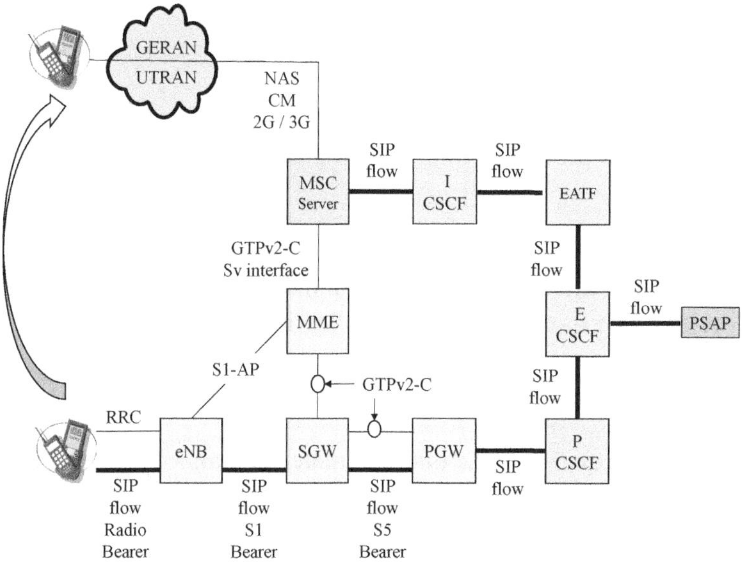

Figure 9.11. *Functional architecture for emergency call control plane*

SIP flow is transmitted to the public safety answering point (PSAP) via the interconnect border control function (IBCF) in the case of IMS interconnection or via breakout gateway control function (BGCF) and MGCF entity in the case of CS interconnection.

The RTP stream is transmitted to the PSAP emergency center via TrGW in the case of IMS interconnection or via IMS GW in the case of CS interconnection.

In the case of IMS interconnection, the IBCF entity ensures the anchor point of the SIP stream.

In the case of IMS interconnection, the TrGW entity ensures the anchor point of the RTP stream.

In the case CS interconnection, the MGCF entity ensures the anchor point of the SIP stream.

In case of CS interconnection, the entity IMS GW entity ensures the anchor point of the RTP stream.

9.2.2. *Procedures*

9.2.2.1. *Registration*

The registration procedure implementing the e-SRVCC function is shown in Figure 9.12.

1) The ATCF entity transfers the SIP REGISTER request to the I-CSCF entity by inserting the `Path` header containing the PATH URI identity and the header `Feature-Caps` with the following parameters:

– `+g.3gpp.atcf`: this parameter informs the session transfer number for SRVCC (STN-SR);

– `+g.3gpp.atcf-mgmt`: this parameter contains the URI identity of the ATCF entity;

– `+g.3gpp.atcf-path`: this parameter contains the identity PATH URI identity of the ATCF entity;

– `+g.3gpp.srvcc-alerting`: this parameter indicates that the MSC Server entities support the SRVCC function during the ringing phase.

```
REGISTER sip:home1.net SIP/2.0
Feature-Caps: *;+g.3gpp.atcf="<tel:+1-237-888-9999>";
+g.3gpp.atcf-mgmt-uri = "<sip:atcf.HomeA.net>";
+g.3gpp.atcf-path="<sip:termsdgfdfwe@atcf.HomeA.net>";
+g.3gpp.mid-call;
+g.3gpp.srvcc-alerting
Path:<sip:termsdgfdfwe@atcf.HomeA.net>,<sip:aga2gfgf@pcs
cf1.HomeA.net:5070;ob>
...
```

2) In the second registration phase, when the mobile has been authenticated, the ATCF entity includes in the message SIP `200 OK`, the Header `Feature-Caps` with the parameter `+g.3gpp.atcf` containing STN-SR number.

```
SIP/2.0 200 OK
Feature-Caps: *;+g.3gpp.atcf="<tel:+1-237-888-9999>"
...
```

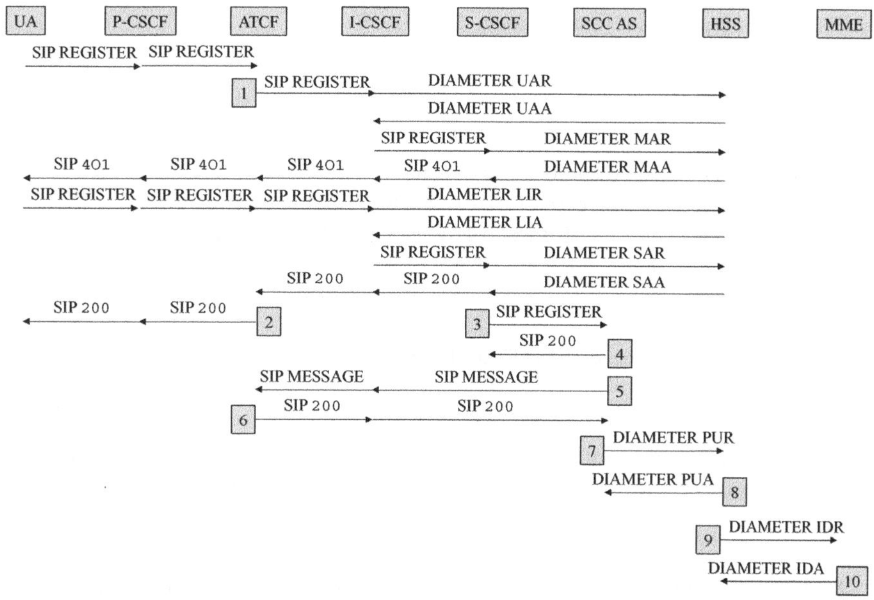

Figure 9.12. *Registration for service continuity*

3) At the end of the mobile's registration, the SCC AS entity receives from the S-CSCF entity the SIP REGISTER message containing the following information created by the ATCF entity:

– the SIP URI identity of the ATCF entity;

– the PATH URI identity for the identification of the mobile while registering;

– the STN-SR session transfer number.

4) The SCC AS entity responds with SIP `200 OK` message to the SIP REGISTER request.

5) The SCC AS entity communicates to the SIP URI of the ATCF entity, in the SIP MESSAGE request, additional information concerning the PATH URI identity: its own access transfer update – transfer session identifier (ATU-STI) and the mobile subscriber ISDN number (MSISDN).

```
MESSAGE sip:atcf.HomeA.net SIP/2.0
...
<?xml version="1.0" encoding="UTF-8"?>
<SRVCC-infos>
  <SRVCC-info ATCF-Path-
URI="sip:termsdgfdfwe@atcf.HomeA.net">
    <ATU-STI>sip:sccas.HomeA.net</ATU-STI>
    <C-MSISDN>tel:+358-50-4821437</C-MSISDN>
  </SRVCC-info>
</SRVCC-infos>
```

6) The ATCF entity responds with the SIP `200 OK` message to the SIP REGISTER request.

7) The SCC AS entity forwards the message DIAMETER profile-update-request (PUR) to the HSS entity to update the STN-SR session number.

8) The message DIAMETER profile-update-answer (PUA) acknowledges the received message DIAMETER PUR.

9) The HSS entity sends the message DIAMETER insert-subscriber-data-request (IDR) to the MME entity for the update of the STN-SR session number.

10) The message DIAMETER insert-subscriber-data-answer (IDA) acknowledges the received message DIAMETER IDR.

9.2.2.2. *Session establishment at originating side*

The procedure for establishing the session implementing e-SRVCC function for an outgoing call is given in Figure 9.13.

1) The establishment of the session is started by the SIP INVITE request transmitted by Alice's UA entity to Bob's UA entity.

The body of the SDP message contains the characteristics of the RTP stream of the offer of Alice's UA entity:

– IPv4 address (192.1.1.1);

– port number (3456).

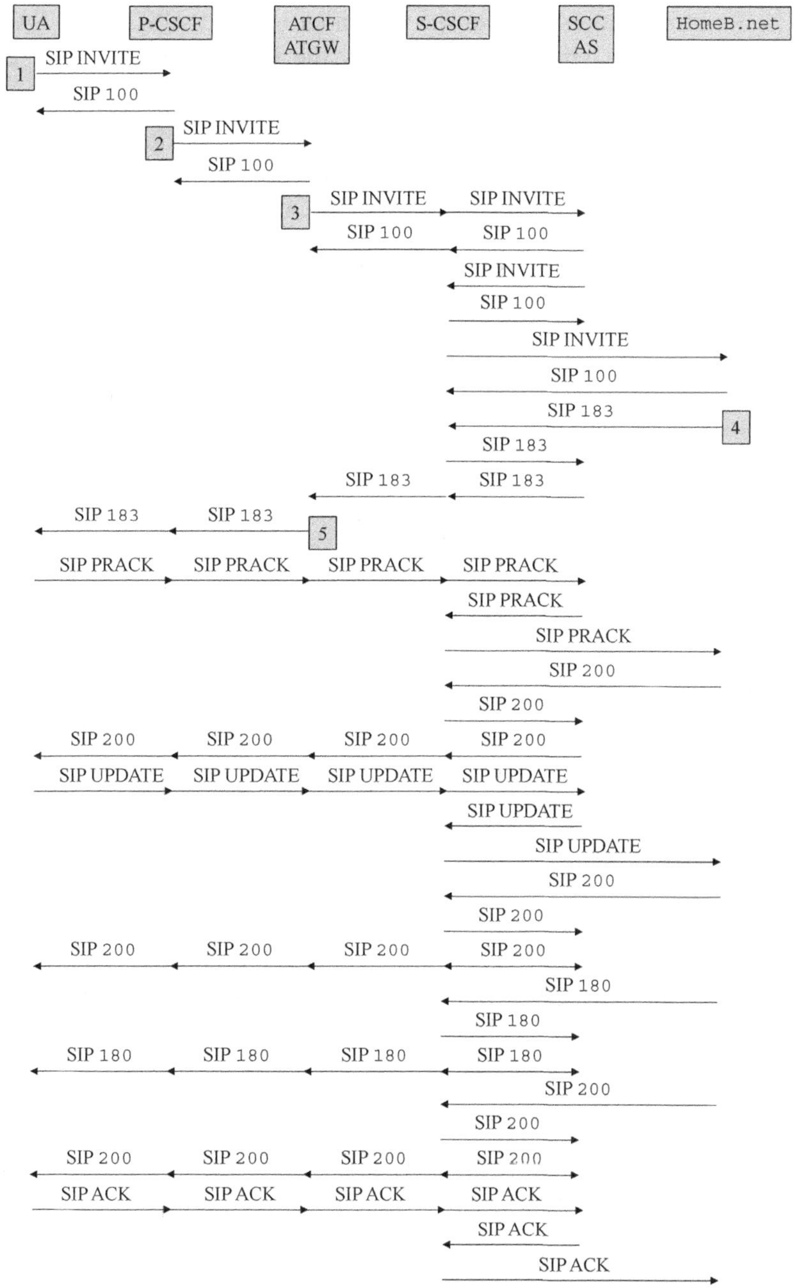

Figure 9.13. *Session establishment for service continuity originating side*

The `Route` headers contain the identities of the P-CSCF and S-CSCF entities, Alice's UA entity does not know the identity of the ATCF entity which, however, is located between the P-CSCF and S-CSCF entities.

2) As the SIP `200 OK` response to the SIP REGISTER request contains the header `Feature-Caps` with parameter `+g.3gpp.atcf`, the P-CSCF entity forwards the SIP INVITE request to the ATCF entity.

3) Upon receipt of the SIP INVITE message, the ATCF entity reserves the resource with the ATGW entity providing the anchor point for the RTP stream.

The ATCF entity transfers to the S-CSCF entity the SIP INVITE request whose associated SDP message, sent to the domain `HomeB.net`, is that of ATGW entity to replace that of Alice's UA entity:

– Alice's IP address (192.1.1.1) is replaced by the value communicated by the ATGW entity (200.1.1.2);

– the port number (3456) is replaced by the value communicated by the ATGW entity (8899).

4) The SIP `183 Session Progress` response from the domain `HomeB.net` contains the characteristics of the RTP stream in the SDP message body:

– IPv4 address (200.2.2.2);

– port number (6544).

5) Upon receipt of the provisional SIP `183 Session Progress` response, the ATCF entity configures the ATGW entity and replaces the SDP message body received from the domain `HomeB.net` by that received from the ATGW entity:

– the IP address (200.2.2.2) is replaced by the ATGW entity address (200.1.1.1);

– the port number (6544) is replaced by the value communicated by the ATGW entity (11234).

9.2.2.3. *Session establishment at terminating side*

The procedure for establishing the session implementing e-SRVCC function for an incoming call is illustrated in Figure 9.14.

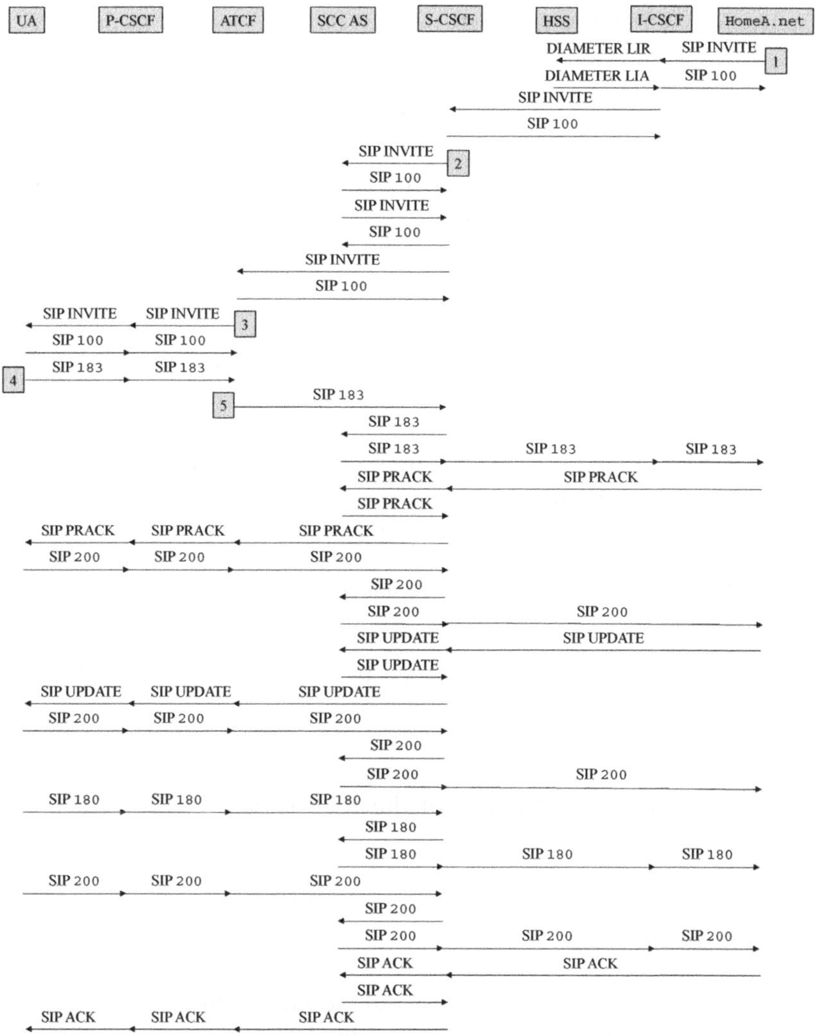

Figure 9.14. *Session establishment for service continuity terminating side*

1) The SIP INVITE request received from the domain `HomeA.net` by the I-CSCF entity contains the SDP message body with the RTP stream characteristics:

– IPv4 address (200.1.1.2);

– port number (8899).

2) The S-CSCF entity forwards the SIP INVITE request to the SCC AS, ATCF, P-CSCF and Bob's UA entities:

– the identity of the SCC AS entity is obtained following the review of the initial filter criteria (iFC);

– the identities of the ATCF and P-CSCF entities are retrieved from the `Path` header of the SIP REGISTER request;

– the IP address of Bob's UA entity results from registration.

3) Upon receipt of the SIP INVITE message, the ATCF entity reserves the resource with the ATGW entity providing the anchor point for the RTP stream.

The ATCF entity transfers to the P-CSCF entity the SIP INVITE request whose associated SDP message, sent to Bob's UA entity, is that of the ATGW entity to replace the one received from the domain `HomeA.net`:

– the IP address (200.1.1.2) is replaced by that of the ATGW entity (200.2.2.1);

– the port number (8899) is replaced by the value communicated by the ATGW entity (10124).

4) The SIP `183 Session Progress` response from Bob's UA entity contains the characteristics of the RTP stream in the SDP message body:

– IPv4 address (192.2.2.1);

– port number (4528).

5) Upon receipt of the provisional response SIP `183 Session Progress`, the ATCF entity configures the ATGW entity and replaces the SDP message body received from Bob's UA entity by the one received from the ATGW entity:

– the IP address (192.2.2.1) is replaced by that of the ATGW entity (200.2.2.2);

– the port number (4528) is replaced by the value communicated by the ATGW entity (6544).

9.2.2.4. *PS-CS inter-system handover*

The PS-CS inter-system handover procedure is given in Figure 9.15.

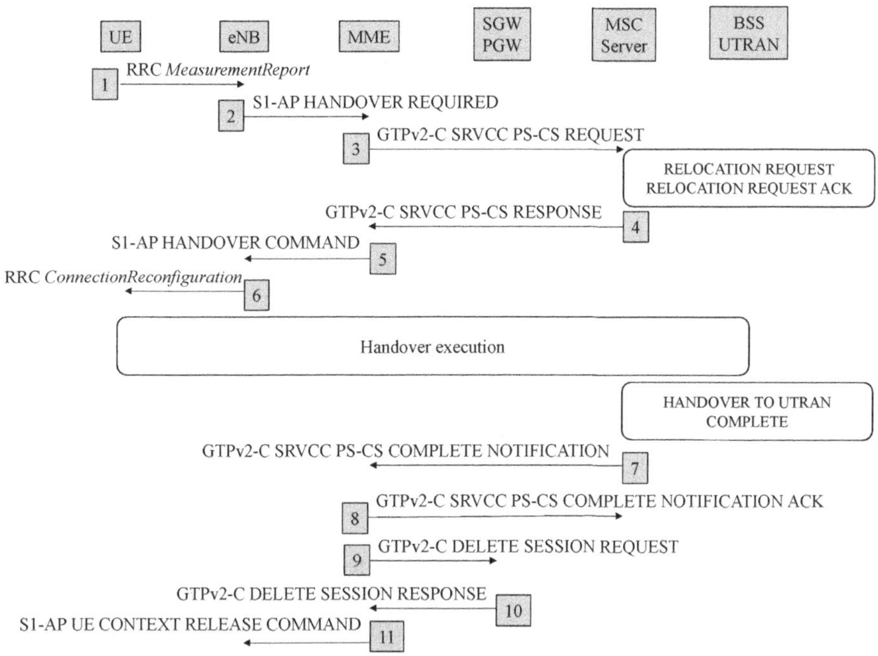

Figure 9.15. *PS-CS inter-system handover*

1) The decision to perform a handover from the EPS network in PS mode to the UMTS or GSM network in CS mode is taken by the eNB on which the mobile is connected.

This decision is taken based on measurement reports from the mobile in the message RRC *MeasurementReport*.

2) The eNB entity sends to the MME entity the message S1-AP HANDOVER REQUIRED. The handover request may also relate to the flow transferred to the 3G UMTS network in PS mode.

3) From the quality class index (QCI), the MME entity separates the audio stream (QCI = 1) and possibly video (QCI = 2) from other flows and initializes them to be relocated either to the MSC server entity (RTP stream) or to the SGSN entity.

The MME entity initiates the procedure of PS-CS inter-system handover by transmitting the message GTPv2-C SRVCC PS-CS REQUEST to the MSC server entity, which contains the STN-SR session number and the MSISDN phone number.

The MSC server entity sends a handover request to the UMTS terrestrial radio access network (UTRAN) of the UMTS network or the base station sub-system (BSS) of the GSM network.

After reserving the resources, the access network acknowledges the request received from the MSC server entity. The answer carries a container that will be transmitted to the mobile. This container contains the radio characteristics of the 2G or 3G cell, which will optimize the mobile connection time.

4) The MSC server entity responds to the MME entity by indicating that resources are available for the execution of the handover in the message GTPv2-C SRVCC PS-CS RESPONSE.

5) The MME entity triggers the execution of the handover by sending the message S1-AP HANDOVER COMMAND to the eNB entity.

6) The eNB entity transmits to the mobile the message RRC *ConnectionReconfiguration*, causing the connection establishment of the mobile to the radio station of the BSS or UTRAN access network.

7) When the mobile is connected, the access network notifies the MSC server entity which informs the MME entity to the completion of handover in a message GTPv2 SRVCC C-PS-CS COMPLETE NOTIFICATION.

8) This message is acknowledged by a message GTPv2-C SRVCC PS-CS COMPLETE NOTIFICATION ACK.

The MSC server entity assigns a temporary mobile subscriber identity (T-TMSI) to the mobile and updates the location of the mobile to the HSS entity.

9) The MME entity induces the deletion of the context of audio and video streams at the SGW by sending the message GTPv2-C DELETE SESSION REQUEST, if the PS-PS inter-system handover is not established.

10) The SGW and PGW entities carry the message received by the message GTPv2-C DELETE SESSION RESPONSE.

11) The MME entity causes the deletion of the contexts of audio and video streams at the entity eNB by sending the message S1-AP UE CONTEXT RELEASE COMMAND to the eNB entity.

After the PS-CS inter-system handover, the CS flow is established between Alice's mobile and the MSC GW entity. The next phase involves the establishment of RTP flows between the MSCGW and ATGW entities.

9.2.2.5. *Access transfer*

The access transfer procedure is described in Figure 8.16.

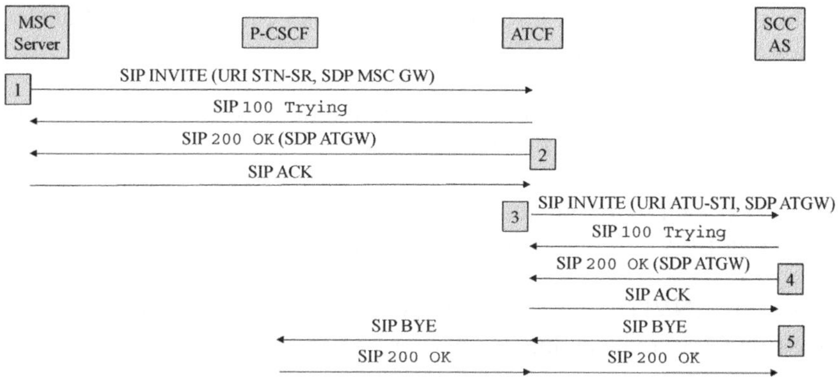

Figure 9.16. *Access transfer for service continuity*

1) The MSC server entity sends the SIP INVITE request to the ATCF entity. The URI identity of the query contains the STN-SR session number mentioned by the MME in the procedure of the PS-CS inter-system handover.

The header `P-Asserted-Identity` contains Alice's phone number (MSISDN).

```
INVITE tel:+1-237-888-9999 SIP/2.0
...
P-Asserted-Identity: <tel:+358-50-4821437>
...
```

The body of the SDP message contains the characteristics of the RTP stream provided by the MSC GW entity:

– IPv4 address (196.1.1.1);

– port number (7236).

2) Upon receipt of the SIP INVITE message, the ATCF entity transfers the SDP message of the MSC GW entity to the ATGW entity and responds with the SIP `200 OK` message containing the SDP message with the RTP flow characteristics provided by the ATGW entity:

– IPv4 address (200.1.1.1);

– port number (5238).

The next phase aims at informing the SCC AS entity about the transfer of Alice's communication to a CS network.

3) The ATCF entity establishes a new dialogue with the SCC AS entity by sending a new SIP INVITE request.

The URI identity of the query contains the ATU-STI identity of the entity SCC AS entity recovered during registration.

The SIP INVITE message incorporates a `Require` header with the value `tdialog` to mean that the header `Target-Dialog` is required.

The header `Target-Dialog` specifies that the existing dialogue (`Call-ID` header, `tag` parameters of the headers `From` and `To`) must be connected with the dialogue established by the ATCF entity.

The `P-Asserted-Identity` header contains Alice's phone number (MSISDN).

```
INVITE sip:sccas.HomeA.net SIP/2.0
...
Require: tdialog
Target-Dialog: me03a0s09a2sdfgjkl491777; remote-
tag=774321; local-tag=64727891
P-Asserted-Identity: <tel:+358-50-4821437>
...
```

The body of the SDP message contains the characteristics of the RTP stream provided by the ATGW entity of the domain `HomeA.net` (IPv4 address 200.1.1.2 and port number 8899) while establishing the session.

As the SDP message body has not changed, there is no need to make an update for the domain `HomeB.net`.

4) The SIP `200 OK` response incorporates the SDP message communicated by ATGW entity of the domain `HomeB.net` (IPv4 address 200.2.2.2 and port number 6544) that the SCC AS entity had retained.

5) The SCC AS entity sends the SIP BYE request to complete the initial dialogue in the P-CSCF and ATCF entities and possibly in Alice's UA entity if it has carried out for the SIP flow the inter-system PS-PS handover simultaneously with the PS-CS inter-system handover.

In the case where Alice's UA entity may not respond with a message SIP `200 OK`, the ending of the dialogue takes place at the initiative of the entities concerned network.

9.2.2.6. *Emergency call*

9.2.2.6.1. Session establishment for an emergency call

The procedure for establishing the emergency call by implementing the e-SRVCC function is illustrated in Figure 9.17.

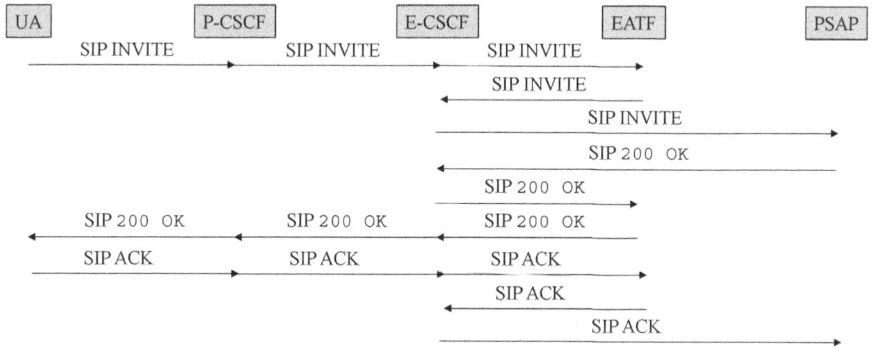

Figure 9.17. *Session establishment for service continuity emergency call*

9.2.2.6.2. Access transfer for an emergency call

The procedure of access transfer for an emergency call is described in Figure 9.18.

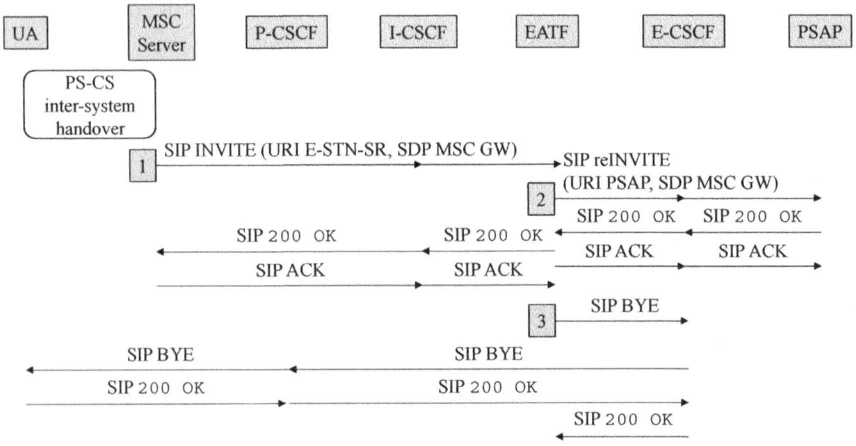

Figure 9.18. *Access transfer for an emergency call*

1) The MSC server entity sends the SIP INVITE request to the EATF entity. The URI identity of the query contains the E-STN-SR number of the EATF entity, configured at the MSC server entity.

The body of the SDP message contains the characteristics of the RTP stream provided by the MSC GW entity.

2) The EATF entity transfers the SDP message received to the entity in charge of the interconnection with the emergency center, in a SIP reINVITE message.

3) The EATF entity issues the SIP BYE request to complete the initial dialogue in the P-CSCF entity, and possibly in Alice's UA entity if it has carried out for the SIP stream the PS-PS inter-system handover simultaneously with the PS-CS inter-system handover.

10

Short Message Service

10.1. Message structure

A short message service (SMS) over the SGsAP protocol allows for a mobile connected to the 4G network to send and receive an SMS in CS (Circuit-Switched) mode.

A short message service over the session initiation protocol (SIP) is a supplementary telephone service offered by the IMS.

The SMS message is generated and interpreted by the short message application layer (SM-AL) between the mobile and the SMS service center (SMS-SC) (Figures 10.1 and 10.2).

The short message transport layer (SM-TL) ensures reliable transmission of SMS messages between the mobile and the SMS-SC entity (Figures 10.1 and 10.2).

The short message relay layer (SM-RL) ensures reliable transmission of SMS messages between, firstly, the mobile, and, secondly:

– the interworking mobile services switching center (SMS-IWMSC) for outgoing SMS messages (Figures 10.1 and 10.2);

– the gateway mobile-services switching center (SMS-GMSC) for incoming SMS messages (Figures 10.1 and 10.2);

The short message control layer (SM-CL) ensures reliable transmission of SMS messages between the mobile and the MSC server entity (Figure 10.1).

For the architecture of SMS over SGsAP, the mobility management entity (MME) ensures relaying SM-CL messages and the MSC server entity that SM-RL messages (Figure 10.1).

For the architecture of SMS over SIP, the IP short message gateway (IP-SM-GW) provides relaying SM-RL messages (Figure 10.1 and 10.2).

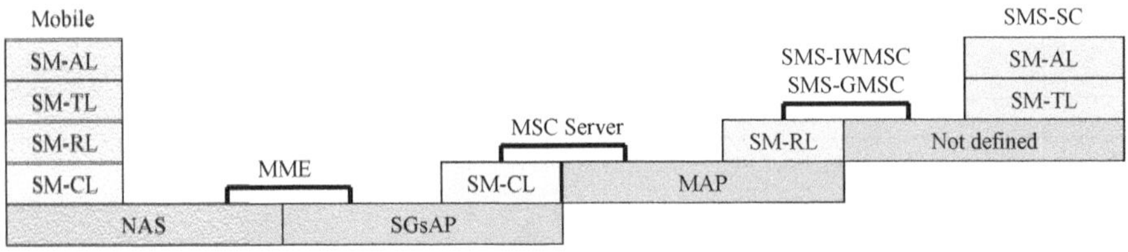

Figure 10.1. *Protocol architecture for SMS over SGsAP*

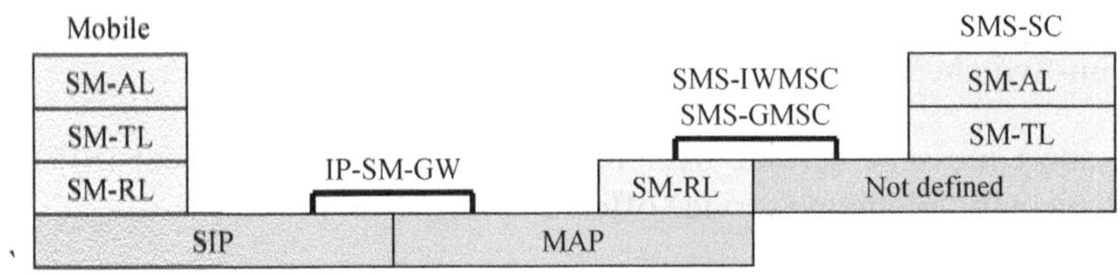

Figure 10.2. *Protocol architecture for SMS over SIP*

10.1.1. *SM-TL layer*

SMS-SUBMIT message carries outgoing SMS messages from the mobile to the SMS-SC entity.

SMS-SUBMIT-REPORT message is an acknowledgment on the part of the SMS-SC entity, positive or negative, of the SMS-SUBMIT message.

SMS-DELIVER message carries incoming SMS messages from the SMS-SC entity to the mobile.

SMS-DELIVER-REPORT message is an acknowledgment on the part of the mobile, positive or negative, of the SMS-DELIVER message or the SMS-STATUS-REPORT message.

SMS-STATUS-REPORT is a report sent by the SMS-SC entity to the mobile which transmits the SMS-SUBMIT message, report relating to the delivery of the SMS message to the destination.

10.1.2. *SM-RL layer*

The header of the SM-RL layer contains the following fields:

– Message Type Indicator: this field indicates the type of message (RP-DATA-ACK RP, RP-ERROR);

– Message Reference: this field contains a sequence number that associates the RP-DATA message and the RP-ACK or RP-ERROR response;

– Originator Address: this field indicates the E.164 phone number of the SMS SC entity issuing the SMS message;

– Destination address: this field indicates the E.164 phone number of the SMS SC entity receiving the SMS message.

The RTP-DATA header contains SMS-SUBMIT, SMS-DELIVER or SMS-STATUS-REPORT messages from the SM-TL layer.

The RP-ACK header is an acknowledgment of the receipt of RP-DATA header and contains the SMS-SUBMIT-REPORT or SMS-DELIVER-REPORT messages from the SM-TL layer.

10.1.3. *SM-CL layer*

The header of the SM-LC layer contains the following fields:

– Protocol Discriminator: this field indicates the nature of the data transmitted. It takes the value “SMS message”;

– Transaction Identifier: this field identifies the number of the transaction;

– Message Type: this field shows the type of message (CP-DATA, CP-ACK, CP-ERROR).

The CP-DATA header contains the headers RP-DATA or RP-ACK and the associated messages of the SM-TL layer.

The CP-ACK header is an acknowledgment of receipt of the message CP-DATA.

10.2. SMS over SGsAP

10.2.1. *Functional architecture*

Short message service over SGsAP is a feature of the circuit-switched fallback (CSFB) mechanism which allows a mobile connected to a 4G network to use the services of the 2G/3G network in CS mode.

The functional architecture of the short message service over SGsAP is described in Figure 10.3.

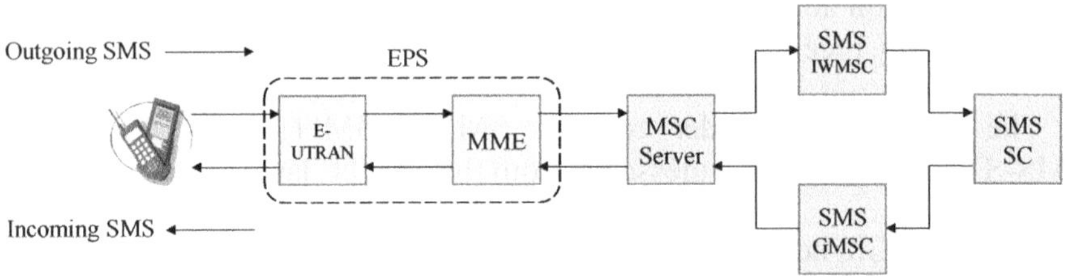

Figure 10.3. *Functional architecture for SMS over SGsAP*

When attaching the mobile, the MME entity attaches the mobile to the MSC server entity, selected by the MME entity.

The MME entity must perform a conversion of the tracking area identity (TAI) of the mobile on the 4G network in a location area identity (LAI) on the 2G or 3G network.

SMS message is conveyed between the mobile and the MSC server entity, and is supported by the following messages:

– non-access stratum (NAS) message between the mobile and the MME entity;

– SGsAP message between the MME and MSC server entities.

SGs interface is the reference point between the MME and MSC server entities for signaling. This interface supports SGsAP signaling.

For an incoming SMS message, the MSC server entity generates a paging to the MME entity.

If the mobile is in standby state on the 4G network, the MME entity triggers the paging procedure for establishing the signaling bearer on LTE-Uu and S1-MME interfaces, allowing the transport of NAS messages.

10.2.2. *Procedures*

10.2.2.1. *Originating side*

The transfer procedure for outgoing SMS messages is given in Figure 10.4.

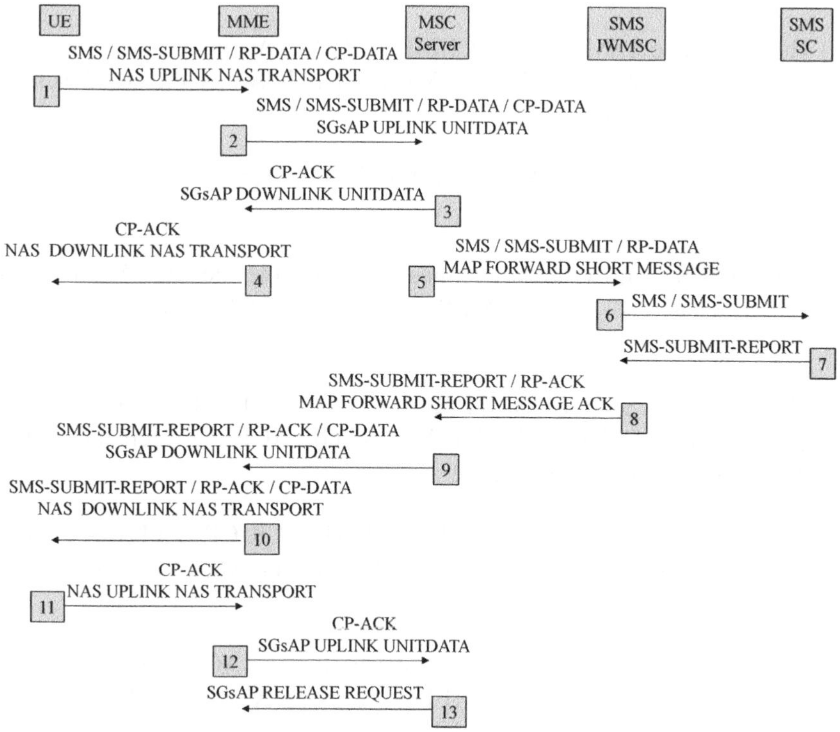

Figure 10.4. *Procedure at originating side for SMS over SGsAP*

1) The message SMS/SMS-SUBMIT/RP-DATA/CP-DATA is carried by the message NAS UPLINK NAS TRANSPORT between the mobile and the MME entity.

2) The message SMS/SMS-SUBMIT/RP-DATA/CP-DATA is conveyed by the message SGsAP UPLINK UNITDAT between the MME and MSC Server entities.

The CP-ACK message is transmitted by the MSC server entity to acknowledge the receipt of the message SMS/SMS-SUBMIT/RP-DATA/CP-DATA.

3) The CP-ACK message is conveyed by the message SGsAP DOWNLINK UNITDATA between the MSC Server and MME entities.

4) The CP-ACK message is conveyed by the message NAS DOWNLINK NAS TRANSPORT between the MME entity and the mobile.

5) The message SMS/SMS-SUBMIT/RP-DATA is carried by the message MAP FORWARD SHORT MESSAGE between the MSC server and SMS-IWMSC entities.

6) The message SMS/ SMS-SUBMIT is transferred to the SMS-SC entity.

7) The SMS SC entity acknowledges receiving the message SMS/SMS-SUBMIT message by sending SMS-SUBMIT-REPORT message to the SMS-IWMSC entity.

8) The message SMS-SUBMIT-REPORT/RP-ACK is carried by the message MAP FORWARD SHORT MESSAGE ACK between the SMS-IWMSC and MSC server entities.

9) The message SMS-SUBMIT-REPORT/RP-ACK/CP-DATA is carried by the message SGsAP DOWNLINK UNITDATA between the MSC server and MME entities.

10) The message SMS-SUBMIT-REPORT/RP-ACK/CP-DATA is carried by the message NAS DOWNLINK NAS TRANSPORT between the MME entity and the mobile.

The CP-ACK message is transmitted by the mobile to acknowledge the receipt of the message SMS-SUBMIT-REPORT/RP-ACK/CP-DATA.

11) The CP-ACK message is conveyed by the message NAS UPLINK NAS TRANSPORT between the mobile and the MME entity.

12) The CP-ACK message is conveyed by the message SGsAP UPLINK UNITDATA between the MME and MSC server entities.

When the SMS-SC entity has the confirmation that the SMS message was delivered to the destination, it sends to the mobile the SMS-STATUS-REPORT/RP-ACK, whose transport is nothing but reproduced steps 7 to 12.

13) The MSC server entity informs the MME entity that the transfer of the SMS message is done by sending the message SGsAP RELEASE REQUEST.

10.2.2.2. *Terminating side*

The transfer procedure for incoming SMS messages is given in Figure 10.5.

1) The SMS-SC entity sends the message SMS/SMS-DELIVER to the SMS-GMSC entity.

2) The SMS-GMSC entity interrogates the home location register (HLR) via the message MAP SEND ROUTING INFO FOR SM to obtain the identity of the MSC server entity.

3) The HLR entity provides the identity of the MSC server entity to the SMS-GMSC entity in the message MAP SEND ROUTING INFO FOR SM ACK.

4) The message SMS/SMS-DELIVER/RP-DATA is transported by the message MAP FORWARD SHORT MESSAGE between the SMS-GMSC and MSC server entities.

5) The message SGsAP PAGING containing the temporary mobile subscriber identity (TMSI) of the mobile, is transmitted to the MME entity.

6) The message S1-AP PAGING, containing the identity shortened TMSI (S-TMSI) is sent to all eNB entities of the TAI area.

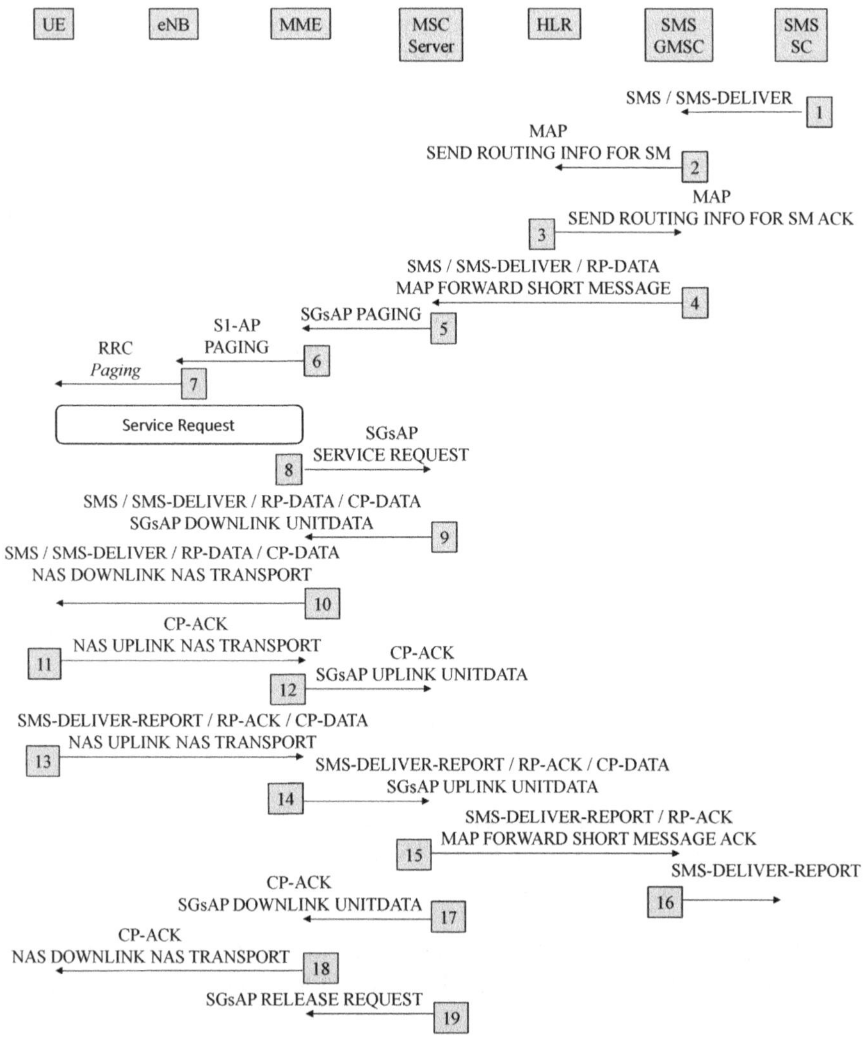

Figure 10.5. *Procedure at terminating side for SMS over SGsAP*

7) The RRC *Paging* message is broadcast by the eNB entities belonging to the TAI area.

The mobile connects to the eNB entity and performs a service request to the MME entity to establish the signaling bearer.

8) The MME entity sends a service request to the MSC server entity, via the message SGsAP SERVICE REQUEST.

9) The message SMS/SMS-DELIVER/RP-DATA/CP-DATA is conveyed by the message SGsAP DOWNLINK UNITDATA between the MSC server and MME entities.

10) The message SMS/SMS-DELIVER/RP-DATA/CP-DATA is transported by the message NAS DOWNLINK NAS TRANSPORT between the MME and the mobile.

The CP-ACK message is transmitted from the mobile to acknowledge the message SMS/SMS-DELIVER/RP-DATA/CP-DATA.

11) The CP-ACK message is conveyed by the message NAS UPLINK NAS TRANSPORT between the mobile and the MME entity.

12) The CP-ACK message is conveyed by the message SGsAP UPLINK UNITDATA between the MME and MSC server entities.

The message SMS-DELIVER-REPORT/RP-ACK is transmitted from the mobile to acknowledge the message SMS/SMS-DELIVER/RP-DATA.

13) The message SMS-DELIVER-REPORT/RP-ACK/CP-DATA is carried by the message NAS UPLINK NAS TRANSPORT, between the mobile and the MME entity.

14) The message SMS-DELIVER-REPORT/RP-ACK/CP-DATA is transported by the message SGsAP UPLINK UNITDATA between the MME and MSC server entities.

15) The message SMS-DELIVER-REPORT/RP-ACK is transported by the message MAP FORWARD SHORT MESSAGE ACK, between the MSC server and SMS-GMSC entities.

16) The SMS-GMSC entity sends the message SMS-DELIVER-REPORT to the SMS-SC entity.

The CP-ACK message is transmitted by the MSC server entity to acknowledge the message SMS-DELIVER-REPORT/RP-ACK/CP-DATA.

17) The CP-ACK message is the message conveyed by SGsAP DOWNLINK UNITDATA between the MSC server and MME entities.

18) The CP-ACK message is conveyed by the message NAS UPLINK NAS TRANSPORT between the MME entity and the mobile.

19) The entity MSC server informs the MME entity that the transfer of the SMS messages is done by sending the message SGsAP RELEASE REQUEST.

10.3. SMS over SIP

10.3.1. *Functional architecture*

The functional architecture of the SMS over SIP is given in Figure 10.6.

The IP-SM-GW entity is an application server of the IMS network.

The IP-SM-GW entity provides the protocol interworking between, firstly, the IMS network and secondly, the SMS-IWMSC entity for outgoing SMS messages or the SMS-GMSC entity for incoming SMS messages.

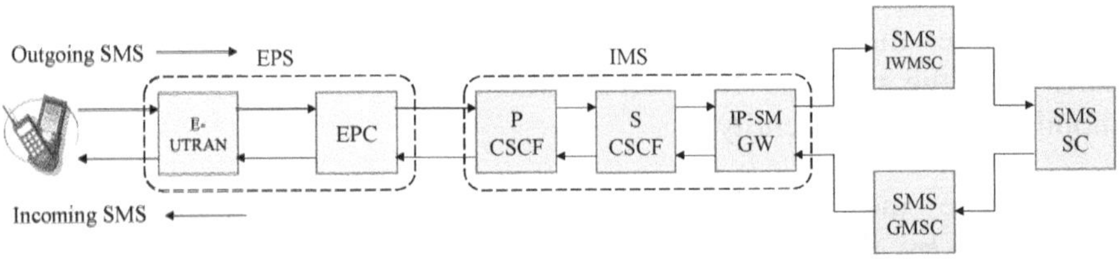

Figure 10.6. *Functional architecture for SMS over SIP*

When the mobile registers with the S-CSCF entity, the S-CSCF entity registers the mobile to the IP-SM-GW entity.

To receive incoming SMS messages, the IP-SM-GW entity registers in turn the mobile to the HLR entity and provides also its address.

The entities SMS-IWMSC, SMS-GMSC and SMSC are not affected by the architecture of SMS over SIP.

10.3.2. *Procedures*

10.3.2.1. *Originating side*

The transfer procedure for outgoing SMS messages is shown in Figure 10.7.

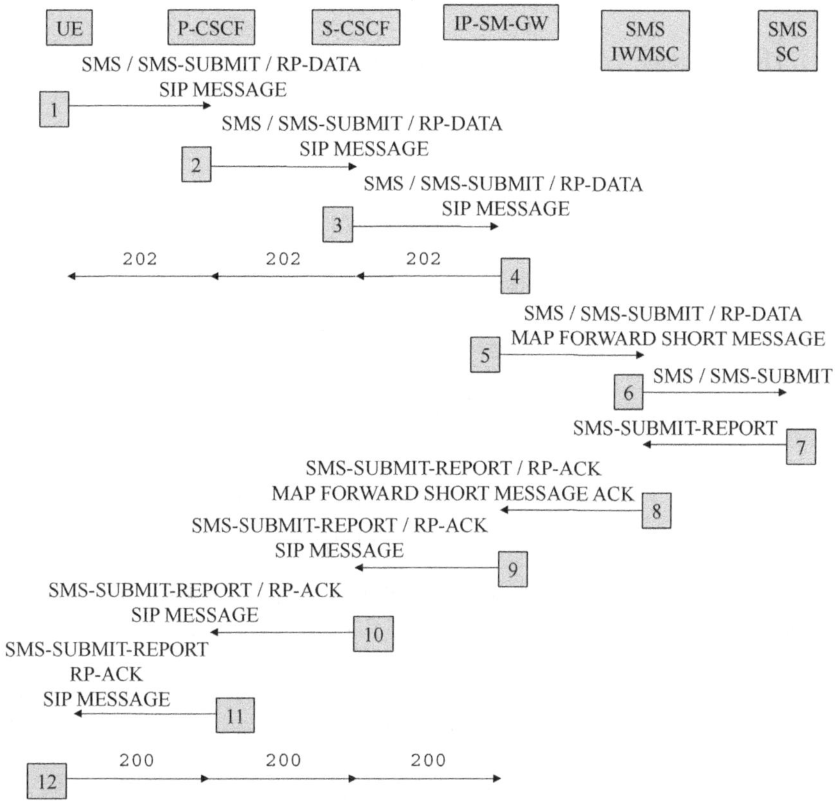

Figure 10.7. *Procedure at originating side for SMS over SIP*

1) The message SMS/SMS-SUBMIT/RP-DATA is carried by the SIP MESSAGE message between the mobile and the proxy call session control function (P-CSCF).

The identity of the request contains the public service identity (PSI) of the SMS-SC entity.

The SIP MESSAGE message indicates in the header `Content Type` that the message body contains an SMS message.

```
MESSAGE sip:sc.homeA.net SIP/2.0
...
Content-Type: application/vnd.3gpp.sms
...
```

2) The message SMS/SMS-SUBMIT/RP-DATA is carried by the SIP MESSAGE message between the P-CSCF and serving CSCF (S-CSCF) entities.

The P-CSCF entity inserts in the header `P-Asserted-Identity` the uniform resource identifier (URI) of the origin of the SMS message.

```
MESSAGE sip:sc.homeA.net SIP/2.0
...
P-Asserted-Identity:<sip:alice@homeA.net>

Content-Type: application/vnd.3gpp.sms
...
```

3) The S-CSCF entity analyzes the received message and the initial filter criteria (iFC) to determine the destination of the SIP MESSAGE message.

The message SMS/SMS-SUBMIT/RP-DATA is carried by the SIP MESSAGE messages between the S-CSCF and IP-SM-GW entities.

4) The answer `202 Accepted` acknowledges the SIP MESSAGE message received by the IP-SM-GW entity.

5) to 8) The messages are identical to those exchanged for SMS over SGsAP, the IP-SM-GW entity playing the same role as the MSC server entity.

9) The message SMS-SUBMIT-REPORT/RP-ACK is carried by the SIP MESSAG message between the IP-SM-GW and the S-CSCF entities.

The identity of the request contains the URI identity of the mobile originating the SMS message.

```
MESSAGE sip:alice@homeA.net SIP/2.0
...
```

10) The message SMS-SUBMIT-REPORT/RP-ACK is carried by the SIP MESSAGE message between the S-CSCF and the P-CSCF entities.

11) The message SMS-SUBMIT-REPORT/RP-ACK is carried by the SIP MESSAGE message between the P-CSCF entity and the mobile.

12) The `200 OK` response acknowledges the SIP MESSAGE message received by the mobile.

10.3.2.2. *Terminating side*

The transfer procedure for incoming SMS messages is given in Figure 10.8.

1) to 4) The messages are identical to those exchanged for SMS over SGsAP, the IP-SM-GW entity playing the same role as the MSC server entity.

5) The message SMS/SMS-DELIVER/RP-DATA is carried by the SIP MESSAGE message between the IP-SM-GW and S-CSCF entities.

The identity of the request contains the URI identity of the mobile receiving the SMS.

The IP-SM-GW entity inserts in the header `P-Asserted-Identity` its own identity.

```
MESSAGE sip:bob@homeB.net SIP/2.0
...
P-Asserted-Identity: <sip:ipsmgw.homeB.net>
...
```

6) The message SMS/SMS-DELIVER/RP-DATA is carried by the SIP MESSAGE message between the S-CSCF and P-CSCF entities.

7) The message SMS/SMS-DELIVER/RP-DATA is carried by the SIP MESSAGE message between the P-CSCF entity and the mobile.

8) The `200 OK` response acknowledges the SIP MESSAGE message received by the mobile.

9) The message SMS-DELIVER-REPORT/RP-ACK is carried by the SIP MESSAGE message between the mobile and the P-CSCF entity.

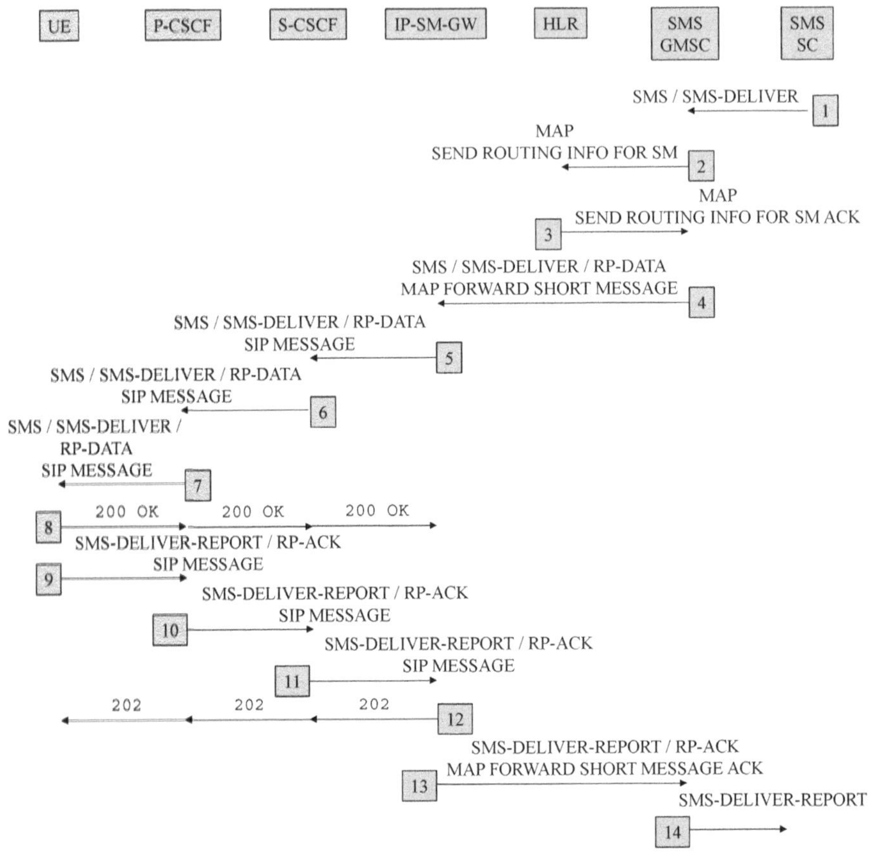

Figure 10.8. *Procedure at terminating side for SMS over SIP*

The identity of the request contains the URI identity of the IP-SM-GW entity.

```
MESSAGE sip:ipsmgw.homeB.net SIP/2.0...
...
```

10) The message SMS-DELIVER-REPORT/RP-ACK is carried by the SIP MESSAGE message between the P-CSCF and S-CSCF entities.

11) The message SMS-DELIVER-REPORT/RP-ACK is carried by the SIP MESSAGE message between the S-CSCF and IP-SM-GW entities.

12) The answer `202 Accepted` acknowledges the SIP MESSAGE message received by the IP-SM-GW entity.

13), 14) The messages are identical to those exchanged for SMS over SGsAP with the IP-SM-GW entity playing the same role as the MSC server entity.

Bibliography

Chapter 1 – Network Architecture

[3GPP TS 23.401] General Packet Radio Service (GPRS) enhancements for Evolved Universal Terrestrial Radio Access Network (E-UTRAN) access

[3GPP TS 23.228] IP Multimedia Subsystem (IMS); Stage 2

[3GPP TS 24.229] IP multimedia call control protocol based on Session Initiation Protocol (SIP) and Session Description Protocol (SDP); Stage 3

[3GPP TS 29.212] Policy and Charging Control (PCC); Reference points

Chapter 2 – Signaling Protocol

[3GPP TS 24.301] Non-Access-Stratum (NAS) protocol for Evolved Packet System (EPS): Stage 3

[3GPP TS 36.331] Radio Resource Control (RRC): Protocol specification

[3GPP TS 36.413] S1 Application Protocol (S1AP)

[3GPP TS 36.423] X2 application protocol (X2AP)

[3GPP TS 29.274] Evolved General Packet Radio Service (GPRS) Tunneling Protocol for Control plane (GTPv2-C); Stage 3

[IETF RFC 3261] SIP: Session Initiation Protocol

[IETF RFC 4566] SDP: Session Description Protocol

[IETF RFC 3428] Session Initiation Protocol (SIP) Extension for Instant Messaging

[IETF RFC 3262] Reliability of Provisional Responses in the Session Initiation Protocol (SIP)

[IETF RFC 3515] The Session Initiation Protocol (SIP) Refer Method

[IETF RFC 6665] SIP-Specific Event Notification

[IETF RFC 3311] The Session Initiation Protocol (SIP) UPDATE Method

[IETF RFC 3588] Diameter Base Protocol

[3GPP TS 29.229] Cx and Dx interfaces based on the Diameter protocol; Protocol details

[3GPP TS 29.329] Sh Interface based on the Diameter protocol; Protocol details

[3GPP TS 29.272] Mobility Management Entity (MME) and Serving GPRS Support Node (SGSN) related interfaces based on Diameter protocol

[3GPP TS 29.212] Policy and Charging Control (PCC); Reference points

[3GPP TS 29.214] Policy and Charging Control over Rx reference point

[3GPP TS 29.215] Policy and Charging Control (PCC) over S9 reference point; Stage 3

[3GPP TS 32.299] Diameter charging applications

Chapter 3 – Basic Procedures

[3GPP TS 23.401] General Packet Radio Service (GPRS) enhancements for Evolved Universal Terrestrial Radio Access Network (E-UTRAN) access

[IETF RFC 3665] Session Initiation Protocol (SIP) Basic Call Flow Examples

[3GPP TS 24.228] Signaling flows for the IP multimedia call control based on Session Initiation Protocol (SIP) and Session Description Protocol (SDP); Stage 3

[3GPP TS 24.930] Signaling flows for the session setup in the IP Multimedia core network Subsystem (IMS) based on Session Initiation Protocol (SIP) and Session Description Protocol (SDP); Stage 3

[3GPP TS 23.167] IP Multimedia Subsystem (IMS) emergency sessions

Chapter 4 – Radio Interface Procedures

[3GPP TS 36.300] Evolved Universal Terrestrial Radio Access (E-UTRA) and Evolved Universal Terrestrial Radio Access Network (E-UTRAN); Overall description; Stage 2

[3GPP TS 36.213] Evolved Universal Terrestrial Radio Access (E-UTRA); Physical layer procedures

Chapter 5 – Service Profiles

[GSMA PRD IR.92] IMS Profile for Voice and SMS

[GSMA PRD IR.94] IMS Profile for Conversational Video Service

[3GPP TS 24.604] Communication Diversion (CDIV) using IP Multimedia (IM) Core Network (CN) subsystem; Protocol specification

[3GPP TS 24.605] Conference (CONF) using IP Multimedia (IM) Core Network (CN) subsystem; Protocol specification

[3GPP TS 24.606] Message Waiting Indication (MWI) using IP Multimedia (IM) Core Network (CN) subsystem; Protocol specification

[3GPP TS 24.607] Originating Identification Presentation (OIP) and Originating Identification Restriction (OIR) using IP Multimedia (IM) Core Network (CN) subsystem; Protocol specification

[3GPP TS 24.608] Terminating Identification Presentation (TIP) and Terminating Identification Restriction (TIR) using IP Multimedia (IM) Core Network (CN) subsystem; Protocol specification

[3GPP TS 24.610] Communication HOLD (HOLD) using IP Multimedia (IM) Core Network (CN) subsystem; Protocol specification

[3GPP TS 24.611] Anonymous Communication Rejection (ACR) and Communication Barring (CB) using IP Multimedia (IM) Core Network (CN) subsystem; Protocol specification

[3GPP TS 24.615] Communication Waiting (CW) using IP Multimedia (IM) Core Network (CN) subsystem; Protocol Specification

[3GPP TS 26.114] Multimedia Telephony; Media handling and interaction

Chapter 6 – Interconnection

[3GPP TS 29.163] Interworking between the IP Multimedia (IM) Core Network (CN) subsystem and Circuit Switched (CS) networks

[3GPP TS 29.165] Interworking between SIP-I based circuit-switched core network and other networks

[3GPP TS 23.205] Bearer-independent circuit-switched core network; Stage 2

[3GPP TS 23.231] SIP-I based circuit-switched core network; Stage 2

[3GPP TS 29.165] Inter-IMS Network to Network Interface (NNI)

Chapter 7 – Handover

[3GPP TS 23.401] General Packet Radio Service (GPRS) enhancements for Evolved Universal Terrestrial Radio Access Network (E-UTRAN) access

Chapter 8 – Roaming

[GSMA PRD IR.65] IMS Roaming and Interworking Guidelines

[GSMA PRD IR.88] LTE Roaming Guidelines

[3GPP TS 29.079] Optimal Media Routing within the IP Multimedia System (IMS); Stage 3

Chapter 9 – Service Centralization and Continuity

[GSMA PRD IR.64] IMS Service Centralization and Continuity Guidelines

[3GPP TS 23.292] IP Multimedia Subsystem (IMS) Centralized Services; Stage 2

[3GPP TS 24.292] IP Multimedia (IM) Core Network (CN) subsystem Centralized Services (ICS); Stage 3

[3GPP TS 23.237] IP Multimedia Subsystem (IMS) Service Continuity; Stage 2

[3GPP TS 24.237] IP Multimedia Subsystem (IMS) Service Continuity; Stage 3

[3GPP TS 23.216] Single Radio Voice Call Continuity (SRVCC); Stage 2

Chapter 10 – Short Message Service

[3GPP TS 23.040] Technical realization of the Short Message Service (SMS)

[3GPP TS 24.011] Point-to-Point (PP) Short Message Service (SMS) support on mobile radio interface

[3GPP TS 23.272] Circuit Switched (CS) fallback in Evolved Packet System (EPS); Stage 2

[3GPP TS 23.204] Support of Short Message Service (SMS) over generic 3GPP Internet Protocol (IP) access; Stage 2

Index

C

D, E

U, V

X

Other titles from

iSTE

in

Networks and Telecommunications

2016

CHIASSERINI Carla Fabiana, GRIBAUDO Marco, MANINI Daniele
Analytical Modeling of Wireless Communication Systems
(Stochastic Models in Computer Science and Telecommunication Networks Set – Volume 1)

2015

BENSLAMA Malek, KIAMOUCHE Wassila, BATATIA Hadj
Connections Management Strategies in Satellite Cellular Networks

BENSLAMA Malek, BATATIA Hadj, BOUCENNA Mohamed Lamine
Ad Hoc Networks Telecommunications and Game Theory

BERTHOU Pascal, BAUDOIN Cédric, GAYRAUD Thierry, GINESTE Matthieu
Satellite and Terrestrial Hybrid Networks

LE RUYET Didier, PISCHELLA Mylène
Digital Communications 1: Source and Channel Coding

PEREZ André
LTE and LTE Advanced: 4G Network Radio Interface

PISCHELLA Mylène, LE RUYET Didier
Digital Communications 2: Digital Modulations

PUJOLLE Guy
Software Networks

2014

ANJUM Bushra, PERROS Harry
Bandwidth Allocation for Video under Quality of Service Constraints

BATTU Daniel
New Telecom Networks: Enterprises and Security

BEN MAHMOUD Mohamed Slim, GUERBER Christophe, LARRIEU Nicolas, PIROVANO Alain, RADZIK José
Aeronautical Air−Ground Data Link Communications

BITAM Salim, MELLOUK Abdelhamid
Bio-inspired Routing Protocols for Vehicular Ad-Hoc Networks

CAMPISTA Miguel Elias Mitre, RUBINSTEIN Marcelo Gonçalves
Advanced Routing Protocols for Wireless Networks

CHETTO Maryline
Real-time Systems Scheduling 1: Fundamentals
Real-time Systems Scheduling 2: Focuses

EXPOSITO Ernesto, DIOP Codé
Smart SOA Platforms in Cloud Computing Architectures

MELLOUK Abdelhamid, CUADRA-SANCHEZ Antonio
Quality of Experience Engineering for Customer Added Value Services

OTEAFY Sharief M.A., HASSANEIN Hossam S.
Dynamic Wireless Sensor Networks

PEREZ André
Network Security

PERRET Etienne
Radio Frequency Identification and Sensors: From RFID to Chipless RFID

REMY Jean-Gabriel, LETAMENDIA Charlotte
LTE Standards
LTE Services

TANWIR Savera, PERROS Harry
VBR Video Traffic Models

VAN METER Rodney
Quantum Networking

XIONG Kaiqi
Resource Optimization and Security for Cloud Services

2013

ASSING Dominique, CALÉ Stéphane
Mobile Access Safety: Beyond BYOD

BEN MAHMOUD Mohamed Slim, LARRIEU Nicolas, PIROVANO Alain
Risk Propagation Assessment for Network Security: Application to Airport Communication Network Design

BEYLOT André-Luc, LABIOD Houda
Vehicular Networks: Models and Algorithms

BRITO Gabriel M., VELLOSO Pedro Braconnot, MORAES Igor M.
Information-Centric Networks: A New Paradigm for the Internet

BERTIN Emmanuel, CRESPI Noël
Architecture and Governance for Communication Services

DEUFF Dominique, COSQUER Mathilde
User-Centered Agile Method

DUARTE Otto Carlos, PUJOLLE Guy
Virtual Networks: Pluralistic Approach for the Next Generation of Internet

FOWLER Scott A., MELLOUK Abdelhamid, YAMADA Naomi
LTE-Advanced DRX Mechanism for Power Saving

JOBERT Sébastien *et al.*
Synchronous Ethernet and IEEE 1588 in Telecoms: Next Generation Synchronization Networks

MELLOUK Abdelhamid, HOCEINI Said, TRAN Hai Anh
Quality-of-Experience for Multimedia: Application to Content Delivery Network Architecture

NAIT-SIDI-MOH Ahmed, BAKHOUYA Mohamed, GABER Jaafar, WACK Maxime
Geopositioning and Mobility

PEREZ André
Voice over LTE: EPS and IMS Networks

2012

AL AGHA Khaldoun
Network Coding

BOUCHET Olivier
Wireless Optical Communications

DECREUSEFOND Laurent, MOYAL Pascal
Stochastic Modeling and Analysis of Telecoms Networks

DUFOUR Jean-Yves
Intelligent Video Surveillance Systems

EXPOSITO Ernesto
Advanced Transport Protocols: Designing the Next Generation

JUMIRA Oswald, ZEADALLY Sherali
Energy Efficiency in Wireless Networks

KRIEF Francine
Green Networking

PEREZ André
Mobile Networks Architecture

2011

BONALD Thomas, FEUILLET Mathieu
Network Performance Analysis

CARBOU Romain, DIAZ Michel, EXPOSITO Ernesto, ROMAN Rodrigo
Digital Home Networking

CHABANNE Hervé, URIEN Pascal, SUSINI Jean-Ferdinand
RFID and the Internet of Things

GARDUNO David, DIAZ Michel
Communicating Systems with UML 2: Modeling and Analysis of Network Protocols

LAHEURTE Jean-Marc
Compact Antennas for Wireless Communications and Terminals: Theory and Design

RÉMY Jean-Gabriel, LETAMENDIA Charlotte
Home Area Networks and IPTV

PALICOT Jacques
Radio Engineering: From Software Radio to Cognitive Radio

PEREZ André
IP, Ethernet and MPLS Networks: Resource and Fault Management

TOUTAIN Laurent, MINABURO Ana
Local Networks and the Internet: From Protocols to Interconnection

2010

CHAOUCHI Hakima
The Internet of Things

FRIKHA Mounir
Ad Hoc Networks: Routing, QoS and Optimization

KRIEF Francine
Communicating Embedded Systems / Network Applications

2009

CHAOUCHI Hakima, MAKNAVICIUS Maryline
Wireless and Mobile Network Security

VIVIER Emmanuelle
Radio Resources Management in WiMAX

2008

CHADUC Jean-Marc, POGOREL Gérard
The Radio Spectrum

GAÏTI Dominique
Autonomic Networks

LABIOD Houda
Wireless Ad Hoc and Sensor Networks

LECOY Pierre
Fiber-optic Communications

MELLOUK Abdelhamid
End-to-End Quality of Service Engineering in Next Generation Heterogeneous Networks

PAGANI Pascal *et al.*
Ultra-wideband Radio Propagation Channel

2007

BENSLIMANE Abderrahim
Multimedia Multicast on the Internet

PUJOLLE Guy
Management, Control and Evolution of IP Networks

SANCHEZ Javier, THIOUNE Mamadou
UMTS

VIVIER Guillaume
Reconfigurable Mobile Radio Systems

Printed and bound by CPI Group (UK) Ltd, Croydon, CR0 4YY
16/09/2026
14963022-0007